THE
INTERNATIONAL
TELECOMMUNICATION
UNION
IN A
CHANGING
WORLD

THE INTERNATIONAL TELECOMMUNICATION UNION IN A CHANGING WORLD

George A. Codding, Jr.
and
Anthony M. Rutkowski

International Standard Book Number: 0-89006-113-0
Library of Congress Catalog Card Number: 81-71049

This book is dedicated with love to

Jennifer

and her generation

and

Kathy

Contents

Preface

The idea for this study evolved from a recognition by the authors of a mutual interest in the International Telecommunication Union, an interest enhanced when we were both present at the ITU's World Administrative Radio Conference held in Geneva in 1979. The project was also promoted by the lack of any current, comprehensive and readable treatise on the ITU and what it does. We decided then to combine for publication our knowledge of and insights into this important international organization which has done much to further the telecommunication revolution.

We owe a debt of gratitude to the many individuals from national telecommunication administrations, the secretariat of the ITU, and private operating agencies who were willing to share with us their thoughts on the ITU, especially those who were kind enough to read and criticize sections of the manuscript. We also wish to express our thanks for the support of the Department of Political Science at the University of Colorado in the preparation of the manuscript, in particular the work of Marilyn Ellis, who did much of the typing. Mr. William M. Bazzy, of Artech House Books, deserves a special word of appreciation for making the publication of this book a priority item.

As regards the coauthorship, Tony Rutkowski assumed primary responsibility for the very important chapters 2, 10, and 11; and George Codding for the remainder, except for the introduction, chapter 1, and chapter 14, for which responsibility was shared. Nevertheless, the work is truly a cooperative project and we thus jointly assume responsibility for any errors of fact, omission, or interpretation.

George A. Codding, Jr.   *Anthony M. Rutkowski*
University of Colorado   Federal Communications Commission*

*The views expressed are those of the author and do not necessarily reflect the views of the Commission.

Introduction

From the time of the introduction of the electric telegraph in the mid-nineteenth century, telecommunication has played an essential role in human activity. Telecommunication quickly became the lifeblood of commerce, government, and the military. Safety of life at sea and in the air was enhanced immeasurably and the average person was entertained, informed, and educated as never before. There are few human activities which were not in some way connected to or influenced by the telecommunication revolution.

This revolution not only continues unabated, it quickens with each passing day. New technologies and new techniques assail us from all sides. New generations of communication satellites and undersea cables make communicating with individuals in other countries as easy as telephoning a neighbor. The future is even more exciting. Optical transmission, dynamic radio systems, mass data storage and processing facilities will soon be linked together and managed by centralized computer intelligence as a vast integrated information network. Human ingenuity in the realm of telecommunication seems to know no limits.

The ultimate use to which technology is put, however, depends upon a number of variables other than the purely technical. Economic, political, and social considerations tend to shape institutional outcomes as much as the technical. Because of the importance of telecommunication to so many critical aspects of national life, in most countries telecommunication was made a government monopoly and in all others it came under strong government control. And when, in response to the needs of commerce and national politics, telecommunication crossed national frontiers, as it did almost from the beginning, it was necessary to find a mechanism through which nations could coordinate international telecommunication for the good of all.

The mechanism that was created for this purpose was the International Telecommunication Union (ITU). Established in Paris in 1865 by representatives from twenty European nations as the International Telegraph Union, the ITU is the oldest of all existing international organizations. Located in Geneva, Switzerland, since 1947, the ITU has a membership of 156 and continues to serve the needs of member nations in the ever-changing, ever-expanding world of telecommunications.

In this book the authors will take a comprehensive look at the ITU from the point of view of its ability to adapt to changing technologies, the changing needs of states, and the changing international political environment in which it operates. The first section is historical in nature, tracing the constant evolution of telecommunication technologies, the problems that their use posed to the international community, and how, through the medium of the ITU, the community went about solving those problems.

The second section will deal with the decision-making apparatus of the ITU as it has evolved, including conferences and meetings of delegates, the Administrative Council, the International Frequency Registration Board (IFRB), the two ITU consultative committees, and the ever-essential Secretariat. Also discussed is the extremely important issue of finances.

In the third section the authors will present a critical analysis of the actual product of the ITU's activities, the tasks that it performs for the various governments that make up its membership. The needs of the older developed nation clientele will be reviewed as well as those of the new, less developed majority.

In a final section, the authors will evaluate the performance of the ITU. An effort will be made to determine whether it still performs functions for the majority of its clientele that are worth the cost of the manpower devoted to its work, as well as the financial contributions which nations make to its expenses. The future directions of telecommunication technology will also be explored, and these developments will be related to the kinds of functions which the ITU may well be asked to perform in the future. The authors will also offer suggestions for change in the ITU's structure and its methods of operation, which they believe will make the ITU better prepared for the next one hundred and fifteen years.

I

History

chapter 1

THE ITU COMES OF AGE (1865-1947)

The International Telegraph Union, parent organization of the ITU, was the first genuine international, intergovernmental organization to see the light of day. As a result, it was forced to achieve on its own a structure and method of functioning that would permit it to meet the problems posed by new communication technologies as well as the demands of its heterogeneous clientele. As such, the history of the ITU is the history of international organizations in general, and the ITU was the model for all such organizations that followed, up to and including the League of Nations.

The group of administrations which began to meet soon after the turn of the century to regulate the new "telegraph without wires" leaned heavily on the Telegraph Union for lessons on how to structure its work.[1] Not only did the new agency emulate the structure of the Telegraph Union, but borrowed its secretariat. For technical and political reasons, the two bodies came into separate existence, but it was only natural that, as the technology of communication progressed, there would be fewer and fewer reasons for them to pursue a separate existence. The decision in 1932 to merge the two and create the International Telecommunication Union was, therefore, a logical and practical move.

The members of the new International Telecommunication Union held only one series of conferences before the world was engulfed by World War II. That fact and the tremendous advances in the use of radio that took place during the war made it imperative to hold a major conference as soon as possible after the cessation of hostilities. By 1947, after some preliminary discussions between the major

---

[1] An actual international organization for radiotelegraph matters was never established. However, the administrations involved worked together in such a way that a *de facto* organization existed. The term International Radiotelegraph Union was often used to describe this working relationship.

powers, that was accomplished. The Atlantic City conferences of 1947 are a landmark in the history of the ITU, since major changes were made in the ITU's structure and method of functioning which remain to the present.

Origins of the International Telegraph Union

At the beginning of the 19th century, Europe was in a state of ferment. The specter of Napoleon I was still abroad in the land, the effects of the industrial revolution were evident everywhere, and the beginnings of a transportation revolution were being signalled by the appearance of the steamship and the railway.

The situation was ideal for the introduction of the new electric telegraph.[2] It was put to use immediately to improve the functioning of railway systems and, as expressed in the ITU's centenary publication, to provide "an efficient means of establishing central control and government . . . "[3] The qualities of the new system of communication rapidly became apparent to news services and other commercial users who had a need for transmitting information rapidly and efficiently. The commercial users were especially welcome, since charges to them could be used to subsidize telegraph use by government and railway. For many years, the least important user of electric telegraph was the average citizen, because of the high cost involved and the low priority given his messages.

All of the users of the growing number of domestic telegraph systems had a strong interest in seeing it extended beyond national borders, so that governments could keep in close contact with embassies and consulates, the railways provide for safe and expeditious international runs, and for use in international trade. Consequently, soon after its introduction on a domestic scale, the electric telegraph began crossing frontiers. However, since each such transaction involved a bilateral treaty, since these transactions were not coordinated, and since there were many states in the small

---

[2] The term "electric telegraph" was adopted to distinguish the new system of telegraph which employed electricity and wires from the older system of visual telegraphy which was in fairly extensive use just prior to the discovery of the electric telegraph.

[3] See ITU, *From Semaphore to Satellite*. Geneva, 1965, p. 29.

geographical area that was Europe, a hodgepodge of arrangements resulted that made the international telegraph service confusing and difficult to use.[4]

The need for some sort of uniformity in various aspects of telegraph
use such as technical standards, codes, and tariffs led certain European countries to collective action. Austria and various German
states came together in a loose grouping known as the Austro-
German Telegraph Union and the Western European states into the
West European Telegraph Union. Although there was some contact
between the two groups and a modicum of effort toward uniformity,
it became obvious that if the telegraph was to attain its true potential, something more was needed.

This was the situation, then, that led Napoleon III to call a conference to meet in Paris in 1865 to draft a treaty that would eliminate
the discrepancies between the practices of the two groups.

The delegates went to the task with good will and within two and a
half months had drafted and signed two documents: the International Telegraph Convention and an annexed Telegraph Regulations. These two instruments accomplished two important things:
1) the creation of an inter-European telegraph network; and 2) the
creation of an international organization to keep the international
network under observation and make changes when deemed necessary.

As regards the international network, in addition to setting forth
the standards for the technology and deciding on a code to be used
—the international Morse code — the delegates to the Paris Conference obtained agreement on such items as the interconnection of
important cities, hours for reception of telegrams, the obligation to
deliver messages from abroad, and the obligation of each country to
dedicate sufficient network capacity to fulfill all requirements for
transmission of messages between telegraph offices with an international connection.

---

[4] The historical material in this chapter, unless otherwise indicated, is from George A.
Codding, Jr., *The International Telecommunication Union: An Experiment in Inter-
national Cooperation.* Leiden: E.J. Brill, 1952. (Reprinted in 1972 by Arno Press, New
York.)

Another area of interest to the delegates at Paris was that of charges for the use of the international network. To foster the use of the international telegraph, the French were of the opinion that all countries represented should establish a "unique" tariff to replace the old system of tariffs by zones. According to the French plan, each country involved would impose only two charges for handling a telegram in its territory, regardless of the distance it travelled: an origin or destination charge (terminal rate), and a charge for handling telegrams by an intermediate telegraph administration (transit rate). As it turned out, after a great deal of discussion, a table of rates was annexed to the regulations, but the charges notified were far from uniform and the larger states obtained the right to add an extra charge for telegrams destined to the most distant parts of their territories.

The other major accomplishment of the Paris Telegraph Conference was the creation of what became known as the International Telegraph Union, the oldest international organization and the direct ancestor of the present-day International Telecommunication Union. The International Telegraph Union was created, in effect, by the provision of the Telegraph Convention which obliged member countries to meet at periodic intervals to keep up with the technical and administrative progress of the international telegraph, the first such meeting to be held in Vienna in 1868, and the decision to allow any country to adhere to the Telegraph Convention by the simple expedient of notifying the French government of its intent through diplomatic channels. The Telegraph Union's original structure was completed at the 1868 Vienna Conference when the decision was made to establish an International Bureau (a secretariat, in modern terms) in Berne, Switzerland.

The International Telegraph Union, 1865 to 1932

In the period from the Paris Telegraph Conference until the decision in 1932 to merge it with the organization created by the International Radiotelegraph Convention in 1932, the International Telegraph Union held numerous meetings throughout Europe to keep the Telegraph Convention and Regulations up to the state-of-the-art of telegraphy and to oversee the rapidly growing international telegraph network.[5]

---

[5] The following is a list of the telegraph conferences held during this period: Vienna, 1868; Rome, 1871-72; St. Petersburg, 1875; London, 1879; Berlin, 1885; Paris, 1890; Budapest, 1896; London, 1903; Lisbon, 1908; Paris, 1925; Brussels, 1928.

During the first part of this period, a great deal of the time of the meetings was devoted to technical problems which surfaced as telegraphy made steady advances. Faster systems were developed, a procedure was discovered which allowed more than a single telegraph message to be sent over the same wire, and the undersea telegraph cable was perfected.

Despite these advances, however, it was soon discovered that the technology associated with the telegraph did not need detailed or extensive regulations. Further, there was a feeling that too strict regulations might hinder the development of that technology. Consequently, administrations were allowed more and more freedom to decide on what technology should be used in their portions of the international network, so long as the service provided was fast and efficient. For example, while the earlier conferences specified the exact type of telegraph apparatus to be used, by 1925 the regulations only provided that:

> Offices between which the exchange of telegrams is continuous or very active are, so far as practicable, connected by direct means of communications reaching the necessary mechanical, electrical and technical standards.[6]

Other major obligations of members of the Telegraph Union were summarized in the same regulations:

> The necessary international channels of communication are to be provided in sufficient number to fulfill all requirements for transmission between offices directly connected.

> The working of these means of communication is arranged by agreement between the administrations concerned.[7]

As telegraph technology assumed a less conspicuous part in the proceedings of telegraph conferences, the problem of rates became paramount. Basically, the problem—which is still with us—is one of fixing a rate for the use of the telegraph that will render the maximum return to the telegraph administration while, at the same time,

---

[6] As quoted in Codding, *The International Telecommunication Union, p. 55.*

[7] *Ibid,* p. 56.

not so high as to encourage the public to use an alternative method of communication. The entire process was complicated by the fact that it was extremely difficult to determine the precise cost of sending a telegram, since many government agencies used the telegraph without charge, and by the fact that in most countries the telegraph was combined with the post offece operation, making it difficult, if not impossible, to separate the costs of the two operations. A somewhat harsh view of the process is found in the report of the American observer to the St. Petersburg Telegraph Conference in 1875:

> The interests of the public who use the telegraph seemed to be entirely subordinated to the interests of the state and to the administrations: that is, to a fear lest any improvement (in the rate structure) might produce less revenue than is got at present, and lest it might throw more work on the telegraph bureau.[8]

Suffice it to say that long and tedious negotiations were the rule in the telegraph conferences before and after World War I over such issues as whether or not to have a common terminal and transit charge for all of Europe, whether non-European and large states should be allowed to charge more, whether to allow telegrams in code and if so how much to charge, the procedure for the settlement of international accounts, and the like.

The only new technological development that occurred in the first thirty-five years of the Telegraph Union was the perfection of the telephone. Although the telephone was adopted for domestic use in some countries in the early 1880s, it was slow to take its place alongside the telegraph for international communications. First, the early telephone was expensive and unreliable, especially over long distances. It did not provide a permanent record of messages sent nor did it bridge language barriers. Second, the telephone was developed at a time when most countries already had an extensive telegraph network and the populace had been educated to use the telegraph as the means of long distance, rapid communication. Third, and equally important, the national telegraph monopolies, which were the rule, were more anxious to protect their investment in the telegraph than to provide the public with a competitive means of communication.

---

[8] *Ibid*, p. 65. .

The subject of telephone regulations was raised by the German government at the 1885 Berlin Telegraph Conference. Because of the opposition to regulation of a number of other European governments, who argued that the telephone service was so new that strict regulations could hinder its development, only a very short addition was made to the telegraph regulations dealing with the telephone. Administrations were given almost complete freedom to interconnect with other telephone services, to determine the technical characteristics of the apparatus to be used, and to fix the charges that were to be made. The only regulation of a substantial nature set the time-unit for charges and a limit to the length of conversations in cases where there were other requests for the use of the line.[9]

Although each subsequent telegraph conference spent some time on the telephone chapter of the regulations, it was to remain in a primitive state almost until the Telegraph Union merged with the organization of Radiotelegraph Convention signatories. It was not until the 1925 Paris Telegraph Conference that a genuine effort was made to make it obligatory for administrations even to interconnect their telephone lines to form a European telephone network, as had been done for the telegraph. By that time the various administrations had adopted such a variety of practices and technology that it was felt it would be impossible to treat the problems in any depth within the confines of a standard telegraph conference. Consequently it agreed to create a Telephone Consultative Committee (discussed in detail later) to do the necessary preliminary work.

On the international organization side, the International Telegraph Union quickly adopted a form that was to last well through World War I. The Paris conference of 1865 laid down the concept of periodic meetings of delegates from administrations to keep the rules and regulations of the international telegraph network up-to-date. At the next conference in Vienna in 1868, it was recognized that a permanent supportive organ was needed to help in the preparation of conferences and to act as an information conduit for the members of the Union concerning telegraph matters affecting the international network. Neutral Switzerland was given the task of installing the new organ (originally called the International Bureau of the Telegraph Administrations, but changed in 1906 to the International

---

[9] Originally five and ten minutes, it was changed to three and six minutes at the Budapest Telegraph Conference in 1896.

Bureau of the Telegraph Union), staffing it, directing its activities, and advancing its operating funds. The expenses of the new Bureau were apportioned to the members of the Union according to a class-unit system.

The one major innovation in the Telegraph Union's organization came in 1925 with the creation of the International Telegraph Consultative Committee and the International Telephone Consultative Committee. As mentioned earlier, by the 1920s it had become apparent that the work necessary to make the telephone a proper partner with the telegraph in the international field was more than could be accomplished by delegates from telegraph administrations in the limited time at their disposal at the regularly scheduled telegraph conferences. In 1924 a group of representatives from post and telegraph administrations met in Paris and decided that the international telephone could benefit from a more or less permanent exchange of views concerning technical and operational problems. To accomplish these objectives, an International Consultative Committee on Long Distance Telephone was created by this group, consisting of annual meetings, a permanent commission, and a secretariat. A Scandinavian proposal at the 1925 Paris Telegraph Conference brought this new organization into a relationship with the International Telegraph Union and charged it with the study of "standards regulating technical and operating questions for international long-distance telephony."[10] A German proposal at the same conference created another consultative committee for the international telegraph. The two committees were given almost complete freedom of organization, with the proviso that any member of the Telegraph Union had the right to participate in the work of these committees and that the two committees were to communicate their recommendations to the Bureau of the Telegraph Union, which would publish them for the benefit of all the members.

The membership of the International Telegraph Union grew by leaps and bounds as additional countries established domestic telegraph and to a lesser extent telephone networks, and connected them with the quickly growing international network. The original twenty became twenty-four by the time of the London conference of 1879 and forty-eight by 1914, including countries as far from Europe

---

[10] See Codding, *The International Telecommunication Union*, p. 36.

as Australia and Argentina. The major absentee was the United States which, although linked to the international network by cable, refused to become a member of the Telegraph Union, contending that since its telegraph and telephone networks had never been nationalized, it could not join an organization that regulated their use. The United States did, however, send observers to meetings.

Two additional aspects of participation in the Telegraph Union are noteworthy. In 1868, the British were invited to attend the Vienna conference on the basis that their government ran the Indian telegraph system. They attended and accepted the final documents in the name of British India. Before the next conference, Britain had nationalized its own domestic telegraph and announced adherence to the Telegraph Convention. Britain then sent two delegations to the Rome Telegraph Conference of 1871-72 and demanded that each be given a vote in the deliberations. "Colonial voting" as the British actions was dubbed, was incorporated in the rules of the Telegraph Union at the St. Petersburg conference of 1875. By 1925 Great Britain, Italy, Portugal, and France all had six additional votes. This practice, which gave an advantage to the colonial powers, was not completely ended until 1973. Participation in work of the Union was also expanded by the Rome Telegraph Conference (1871-72) when ten private telegraph companies were allowed to participate in all meetings and committees without the right to vote.

The 1925 Paris International Telegraph Conference was of special significance for its decision to request the signatories to the International Radiotelegraph Convention, at their next conference, to consider the advisability of a merger. The groundwork for a merger had already been laid in the numerous Inter-Allied meetings which were held between 1919 and 1922, a development that will be discussed in the following section.

Regulation of Radiocommunication, 1903-1932

Radio came into practical application near the end of the 19th century, and its first use was for communication between ships at sea and the land. Early radio tended to work best over water, and there was an important need to be filled in that area. Until the advent of radio, there was no means of communicating with ships once they were out of sight of land except by means of chance encounters with other ships. Continuous contact with elements of the world's merchant fleet was important both for safety of life and for commercial considerations.

There was little surprise, therefore, that radio had an immediate acceptance. One of the first groups to exploit the new discovery commercially was the Marconi Wireless Telegraph Company. On the basis of its advanced technology and commercial adeptness, the Marconi Company began to dominate the new world of radio. Marconi transmitters were soon found strategically located on the world's trade routes, and more and more ships began to carry Marconi radios and Marconi radio operators. Marconi's success was such that its directors evidently concluded that, with the right tactics, they could dominate the entire market. In pursuit of this objective, the Marconi Company ordered its operators to refuse to communicate with operators using other than Marconi equipment.

These steps by the Marconi Company were a step backward for the rapidly growing maritime communication system. Perhaps equally important, they occurred at a time when two other major companies — one in the United States and the other in Germany — had decided to enter the developing, lucrative radio market.

It was, therefore, not surprising that the first radio conference was called by the German government, to meet at Berlin in 1903 for the express pupose of obtaining agreement that all coast stations would in the future accept telegrams from any ship station regardless of the brand of equipment used. Seven of the nine countries attending the 1903 Berlin conference agreed: Austria, France, Germany, Hungary, Russia, Spain, and the United States. Two countries whose agreement was necessary to achieve the purpose of the conference refused: Great Britain and Italy.

This conference, the Preliminary Conference Concerning Wireless Telegraphy, produced a Final Protocol which was to be taken back to the respective governments to be reviewed as a basis for an international agreement. The Protocol dealt not only with communication between ships and shore stations, it also provided that portions of the St. Petersburg Telegraph Convention be applicable to wireless communication, that priority of communication be give to ships at sea, and that a station should operate without causing interference to other stations. In general, the provisions were not applicable to government stations which were not open to public communications.

The German government called on all interested countries to send delegates to attend an international conference in Berlin in 1906 to fashion a binding treaty based on the 1903 protocol. Twenty-nine countries responded and, under the guidance of the Germans, drew up an International Radiotelegraph Convention and an annexed Regulations in the example of the International Telegraph Union. While this gathering came no closer to changing the monopolistic practices of the Marconi Company than its predecessor, it did establish several of the fundamentals of the radio regulations that we know today. The Convention specified that certain bands of radio frequencies be used by specific radio "services." For example, frequencies below 188 kc/s were allocated to long distance communication by coast stations, and frequencies between 188 kc/s and 500 kc/s were allocated to government stations not open to public correspondence. In addition, each nation was obligated to give notice concerning their stations' operations through the International Bureau of the Telegraph Union, and to prevent interference with existing stations. The practical effect of these provisions was to establish a process by which rights to use specific radio channels in prescribed geographical areas were vested on a first-come basis.

On the organizational side, the delegates to the 1906 conference decided to meet at regular intervals to keep up with the rapidly changing technology of radio. The convention would be revised by a plenipotentiary conference and the appended Radio Regulations would be amended by an administrative conference. Rather than create a new secretariat, the International Bureau of the Telegraph Union was asked to carry out such support duties as the 1906 Convention should require. The most important of these administrative tasks were to serve as an official repository for, and to publish information relative to, radio services (especially the list of radio stations mentioned above), and to act as a clearinghouse for proposed amendments to the Convention and Regulations. The Telegraph Union agreed, and the first step took place toward a marriage which was not consummated until 1936.

Only one more international radio conference was held before the beginning of World War I. This conference, held in London in 1912, was significant mainly for the agreement of the Marconi people to end the practice of refusing to communicate with others solely on the basis of the type of equipment being used. The 1912 conference also drew up a new set of regulations concerning the better utilization of radio on ships, largely as a result of the *Titanic* disaster.

During the war, the use and development of radiocommunication proceeded at a rapid pace. Two commissions were formed by the Allied powers to effect coordination of communication services. The work of these commissions set in motion a series of events which eventually resulted in the creation of the International Telecommunication Union. Perhaps the most interesting aspect of these meetings was the identification of fundamental issues and the exploration of various solutions. Although telecommunication technology and the composition of the world community have changed dramatically over the years since those meetings, the questions that were raised and the solutions that were discussed have not.

Near the end of the war, the Inter-Allied Radiotelegraph Commission decided that it should continue its meetings in order to begin preparations for a new international radio convention. The war had not only revealed the utility of telecommunication, but also the necessity for effective international cooperation. After many meetings in 1919, the EU-F-GB-I Protocol of August 25, 1919, was produced by the commission.[11] It set forth a number of possible changes to the existing 1912 London convention, and urged interested nations to plan to meet again in the near future to consider these matters further.

The United States, newly thrust into the forefront of foreign affairs, took the lead and hosted a preliminary conference to do considerably more than modify the old convention. It proposed to create a new Universal Electrical Communications Union " . . . with the object of simplifying communications by bringing all methods of electrical transmission as far as practicable under the same rules."[12] The

---

[11] France, Office of the Minister of War, Cipher Section, *EU-F-GB-I Protocol of 25 August 1919,* Paris, 1919. The letters EU-F-GB-I stood for the names, in French, of the countries participating in the Commission: United States, France, Great Britain and Italy.

[12] Universal Electrical Communications Union, *Draft of Convention and Regulations.* Washington, D.C., December, 1920, Preface.

Allied powers met in Washington at the invitation of the U.S. and in ten weeks produced a draft convention and regulations setting forth basic international institutional features for telecommunication which were eventually used as working documents for the next international radiotelegraph conference held at Washington in 1927. Some of the features included: 1) the placing of all telecommunication under the aegis of a single permanent union; 2) the establishment of a Central Bureau for carrying on continuing administrative activities; 3) the establishment of a council to serve as an interim governing body between formal meetings for the Union; 4) the establishment of an International Radio Technical Committee to serve as a consultative organ; 5) the allocation of frequency bands among defined classes of radio users called "services;" 6) the adoption of various schemes for vesting rights to use specific radio channels free from harmful interference from stations of other states; and 7) the recognition that private enterprises may deal directly with the Central Bureau and with administrations concerning their telecommunication operations. Thus, the basic structure of the Union and its functions that exist today were devised at the 1920 Washington Preliminary Conference.

Some of the unsolved questions were referred to a Technical Committee on International Radio-Communication which met in Paris for three months in 1921. The most thorny question then, as today, concerned the kind of administrative scheme that should be adopted for vesting rights in frequency usage. Despite dozens of stormy sessions during that summer, a consensus could not be reached.[13] The spirit of post-war international cooperation had waned and no important meetings on the subject were held again until 1927.

In the meantime, significant changes in radio technology and applications had taken place. Much better equipment was being produced, ever higher frequencies were being employed, and broadcasting had made its debut. While the state of the art was changing so rapidly that government officials and private industry were apprehensive about any kind of definitive standards being establish by international bodies, the rush of many nations and entrepeneurs to establish broadcasting stations called for some kind of international agreement.

---

[13] See [Report of the] *Technical Committee on International Radio-Communication,* Paris, 1921, pp. 30-33.

The broadcasting situation had become difficult by the mid-1920s in Europe and in 1925 the British Broadcasting Company (BBC) called together a number of European administrations to establish a means of cooperating in the use of radio broadcasting. The meeting produced the International Broadcasting Union (U.I.R.). One of its main tasks was to consider the problem of interference between European stations. For the next twenty years, the U.I.R. played a key role in fashioning European broadcasting plans which were incorporated in ITU agreements by reference. Eventually, many of the U. I. R. responsibilities were also merged into the ITU.

When the nations of the world met at Washington in 1927, they finally recognized the need for comprehensive radio regulations. Many of the radio-related features devised at the 1920 Washington preliminary conference were added to the old 1912 London convention, and the resulting "International Radiotelegraph Convention, and General and Supplementary Regulations" was signed by all the major nations of the world except Russia. Although the U.S.S.R. submitted proposals, it was not allowed to attend because of western antipathy toward the new government.

The Supplementary Regulations (later called the Additional Radio Regulations) consisted of a separate body of articles dealing with charges to be imposed when international mobile communication services were provided. As we have noted, the United States had refused to become a member of the International Telegraph Union because U.S. telegraph and telephone services were provided by private enterprise. In creating the Radio Regulations, the Berlin conference of 1906 had borrowed certain technical and administrative articles from the Telegraph Convention of 1875 and applied them to international radiocommunication. The London conference of 1912 followed suit, but added a number of provision concerning rates. The United States did not object to the London action itself, but made a formal reservation to all of the provisions in question. At the 1927 Washington conference, it was decided to place all articles dealing with radio charges in a supplemental section so that the objecting nations—namely the U.S. and Canada—could adhere to all of the general radio regulations but, by refusing to ratify the Supplemental Regulations, could avoid being bound by provisions they felt were impossible to apply.

The important and recurring issue of phasing out obsolescent technology was also considered at the 1927 conference. For many years, spark-type radio transmitters had been used by every ship on the high seas. Though relatively cheap, such transmitters occupied a broad frequency bandwidth. After World War I vacuum tube transmitters became available and were rapidly employed because they used much less of the radio spectrum. Despite complaints of ship owners that extreme financial losses would be incurred by eliminating the old spark transmitters, the conference decided to establish a timetable for the elimination of all such transmitters. Here, in a very definitive way, the economic value of the radio spectrum first became obvious.

The single significant organizational change produced by the 1927 conference was the creation of the International Technical Consultative Committee for Radioelectric Communications. The concept of such a body had been advanced as early as the meeting of the Inter-Allied Radiograph Commission at Paris in 1919. The Telegraph Union had already created separate Telegraph and Telephone Consultative Committees in 1925. The creation of an analogous Radio Committee became the subject of considerable controversy at the 1927 conference. Opposition came from several quarters: the United States argued that such a body might hinder the evolution of radio by adopting premature standards, the French were afraid that some private companies might obtain unfair advantage over others by gaining approval of their technology, and the British argued that it was unnecessary since no binding decision could be made between conferences anyway. A concensus was reached when the decision was made to confine the work of the committee exclusively to technical and operational matters. As we shall see, this organ is now probably one of the most valuable in the ITU for precisely those countries that argued against its creation.

The 1927 conference also took up the request made at the 1925 Paris Telegraph Conference that it consider the possibility of unifying the separate Radiotelegraph Convention and the Telegraph Convention into a single international instrument. Almost without opposition, the 1927 Washington conference agreed that such a move would eliminate much unnecessary duplication and would provide for better collaboration on matters of common concern. In order to effect this union, it was decided to hold the next meeting at Madrid in 1932, at the same time and place as the next meeting of the International Telegraph Union.

In the period between 1903 and 1932, the basic international institutional arrangements for radiotelegraph communication were conceived and implemented. First, all interested nations would meet from time to time and draw up the various technical and operating standards which are necessary for communication to occur. One of the most important of these standards is the Table of Allocations which specifies which radio "services" are to use which frequency bands both universally and in various parts of the world. Second, a permanent, central administrative bureau would be utilized to collect, record, and publish certain information pertaining to telecommunication. Third, through various mechanisms a nation could acquire a right for its radio stations to use specific radio channels without harmful interference from the stations of another nation.

From Merger to Atlantic City, 1947

The 13th International Telegraph Conference and the 4th International Radiotelegraph Conference met simultaneously in Madrid in 1932 to work out the details of a merger. Eighty countries and sixty-two private companies and international organizations were represented at the Telegraph Conference, and sixty-five countries and sixty-four private companies and international organizations at the Radio Conference. Ten joint meetings of the Plenary Assemblies were held, and two joint committees were created for consideration of common questions.

Merger of the two organs was accomplished through the creation of a single convention containing the major elements of both the Telegraph and Radio Conventions to which were attached the respective Telegraph, Telephone, Radio, and Additional Radio Regulations. After token opposition from those who wanted to keep the word "telegraph" in the name out of respect for the older of the two bodies, the name International Telecommunication Union was chosen and the word "telecommunication" was defined as including "[a]ny telegraph or telephone communication of signs, signals, writings, images, and sounds of any nature, by wire, radio, or other systems or processes of electrical or visual (semaphore) signaling."[14]

---

[14] Codding, *The International Telecommunication Union p. 140.*

In order to assure the cooperation of certain countries, especially the United States, the new combined International Telecommunication Convention was confined to generalities concerning the various services and organizational details, and to the usual housekeeping articles of any international treaty. Anything of a controversial nature was confined to the respective regulations, and members of the new Union were given the right to choose which of the regulations they would be bound by.[15]

Although the main accomplishment of the Madrid conference was the creation of a single international organization in the field of communication, a great deal of work was done at the same time on the service regulations. While the telegraph people were occupied by the problem of a common monetary unit, necessitated by financial problems accompanying the economic disruptions of the great depression, and charges for code telegrams, the telephone people were arguing over charges for telephone messages in and outside of Europe.

The radio people were also facing difficulties in finding a place in the radio spectrum for new services. The use of radio for both aeronautical mobile communication and broadcasting had come into increasingly extensive use in the late 1920s, and a place had to be found for them in the frequency allocation table. The maritime interests, however, did not want any such frequencies to be taken from their portions of the spectrum. Fortunately, because of the nature of propagation characteristics of the contested frequencies, a compromise could be reached whereby the low and medium bands were divided into a European region and "other regions." This permitted the Europeans to apportion more frequencies to broadcasting and the United States and Canada to continue operation as before. The Madrid conference continued the process of notifying frequencies to the Berne Bureau to obtain rights to be free from harmful interference, and having the Bureau publish them.

The sole activity of the new International Telecommunication Union prior to the outbreak of World War II was the meeting of a combined Telegraph and Telephone Conference and Radio Conference in Cairo in the inauspicious year of 1938. As had occurred at Madrid, joint meetings were held to treat common questions such as

---

[15] Article 2 of the new Convention, however, provided that each signatory was obliged to sign at least one of the three major Regulations.

voting rights and languages to be used in the conference debates, while all other problems were dealt with in meetings confined to the respective specialists. For the radio experts, the major problem—as had been the case at Madrid—was the allocation of frequencies to services. A band between 6,500 and 23,380 kHz was allocated to intercontinental air routes. Broadcasting picked up some frequencies in the 2,300 to 4,900 kHz and 6,000 to 25,000 kHz bands at the expense of the fixed and mobile services, and two allocations to broadcasting were made in the high frequency range despite the opposition of the United States.[16] Some new frequencies were given to the amateurs, although not as many as desired by the United States. And because their propagation characteristics were of a limited range, it was possible to prevent a major dispute by dividing some of the frequencies in the 25 MHz to 200 MHz range into a European Region and "other regions." The Europeans were thus allowed to assign portions of the frequency band to specified services, and the U.S. was free to be flexible in the use of the entire band.

The Cairo conference also made some important changes in the operation of the consultative committees. The Radio Consultative Committee's function was expanded from the study of technical radio questions to the study of technical questions *and* operating questions, and the Telegraph Consultative Committee was henceforth permitted to study operating questions and rate questions in addition to technical questions.

While the World War II years were quiet ones for the ITU — the Bureau continued to operate in Berne with a reduced staff — radio technology evolved substantially. Radio was used by almost every branch of the military service during the war, with all of the related research and development that such usage entails. Radar evolved, as did a new method for establishing the position of ships at sea with considerable accuracy, known as loran. Broadcasting was used extensively both for internal communication and for overseas propaganda broadcasts.

Radiocommunication was also important in the immediate postwar period for purposes of reconstruction and development and for

---

[16] This decision gave rise to a U.S. reservation which read: "The United States of America reserves for itself the right to use the frequency band 21, 650 to 21, 750 kc/s for mobile as well as broadcasting services." See Codding, *The International Telecommunication Union,* pp. 165-66.

getting the messages of the major powers across to others. All this activity produced a constant expansion and sophistication of radio technology and its use, which meant that developing international radio agreements would have a high priority for the Union in post-war conferences.

The Atlantic City Conferences, 1947

The three Atlantic City Conferences of 1947, the plenipotentiary, the administrative radio conference, and the high frequency broadcasting conference, are worthy of special note because of the many significant changes that were made to the ITU and its functions. These included changes in membership and voting, the creation of new components within the Union, and the adoption of drastically different approaches to the management of the radio frequency spectrum. Many of these changes were so drastic that more than ten years of additional meetings were required to resolve all of the issues that were raised. The details of these subsequent meetings are discussed in Chapter 2.

As was the case in a number of other areas of international endeavour — food, health, and civil aviation, to name a few — the United States, which was relatively unscathed by the war, initiated the post-war collaboration in the field of telecommunication. First, a preparatory conference was held by the allied powers in Moscow in 1946 at the invitation of the Russian government to prepare the way for a renewal of ITU activities. The suggestion for the conference, however, had come from the United States after consultation with the other World War II allies. In many striking respects, history was repeating itself. The impetus for the conferences, the participant countries, the leadership roles, and the approaches taken were almost the same as those immediately following World War I.

Shortly after the Moscow Preliminary Conference, the United States, in consultation with twenty other members as specified in the Convention, issued an invitation to come to Atlantic City in 1947 to reactivate the ITU and to solve some of the major problems brought on by the rapid expansion of new technologies in radio. Despite a great deal of criticism of the United States for failure to hold the first post-war conference at the home of the ITU in Switzerland, seventy-four countries sent some 600 delegates to attend. Also participating were numerous observers from private companies and international organizations.

The first of the three conferences that occurred at Atlantic City in 1947 was the International Radio Conference which began in May. When general agreement was obtained regarding the kinds of changes to the Radio Regulations which were necessary, the International Telecommunication Conference was convened in July, followed by the International Conference on High Frequency Broadcasting in August.[17] All three conferences wound up their business by the beginning of October.

The International Telecommunication Conference drafted a new ITU Convention which significantly affected the rights and obligations of Union members. Henceforth, only sovereign states could become members. This was a hotly debated item which saw its resolution in compromise; the requirement would only apply to new members. Existing territorial members were not affected. In addition, the simple voting majority was retained, except in the case of admission of a new non-UN member, as the result of the hard-fought efforts of a number of countries, including the United States. As we will see, this, and the ITU tradition of one vote per member, later became advantageous to the new developing member countries.

The telecommunication conference also included a provision in the new convention to the effect that all members would automatically be bound by the administrative regulations. This caused some concern to nations like the United States, which do not closely regulate the operation of their telegraph and telephone systems which are in the hands of private companies, and resulted in a number of reservations regarding the Telephone, Telegraph, and Additional Radio Regulations. The Plenipotentiary Conference was confirmed as the supreme organ of the Union, to meet at regular five year intervals and make needed changes to the Convention, and to provide basic policy guidance.

The decision of the Moscow Preliminary Conference, ratified at Atlantic City, that the ITU should become a specialized agency of the United Nations meant that the Union would have to take on

---

[17] The Radio Conference had jurisdiction over the Radio Regulations, and thus was similar to a general World Administrative Radio Conference (WARC) of today. The Telecommunication Conference had complete jurisdiction over the Telecommunication Convention and the ITU as an organization, and thus was similar to current plenipotentiary conferences. The H.F. Broadcasting Conference had somewhat narrow jurisdiction, and was equivalent to a specialized WARC.

some of the structural characteristics of the other organizations that were also becoming specialized agencies.[18] There would be a need for an official organ made up of government delegates, which would maintain closer contact with the appropriate United Nations organs than could the occasional plenipotentiary conferences. For this purpose, the Atlantic City Telecommunication Conference created an Administrative Council made up of representatives of eighteen members of the Union elected by the Plenipotentiary Conference, to meet annually at the seat of the Union in order to supervise the work of the ITU and its relations with other international organizations in the interim between meetings of the plenipotentiary.

The second spin-off of the proposed United Nations connection involved the Berne Bureau. Until this time, the Secretariat of the ITU (or Berne Bureau) was financed, supervised, and staffed by the Swiss government. The other United Nations agencies being created as a part of the new system, and the U.N. itself, were being given international secretariats made up of nationals from a number of member countries, rather than only one. The successful move at Atlantic City to replace the Bureau by a true international secretariat was motivated by several factors. Some felt that the ITU should have an international secretariat, on principle; others disliked the fact that any one country should have such a hold over the basic organ of an important international organization. Still others wished to remove it from the hands of the Swiss. The new General Secretariat was to be headed by an elected Secretary-General and Deputy Secretary-General. It was to be housed in French-speaking Geneva, Switzerland, and to be progressively internationalized.

As part of the effort to operate the ITU on a more structured and regularized basis, the Radio Consultative Committee (CCIR) was made a continuing, instead of periodically, convening body. The change also reflected the growing importance of CCIR work in establishing international standards, as well as providing advice to the newly created International Frequency Registration Board (IFRB).

---

[18] For a discussion of the hands-off attitude of the early ITU and its predecessors to the League of Nations, see Codding, "The Relationship of the League and the United Nations with the Independent Agencies: A Comparison," *Annales d'Etudes Internationales,* 1970, pp. 1165-87.

One of the potentially most important changes instituted by the Telecommunication Conference was the creation of a new body to perform the specialized administrative and advisory functions required by the new rights vesting mechanism which had been set up by both the Radio and the High Frequency Broadcasting Conferences: the International Frequency Registration Board. Its members were to serve as "custodians of an international public trust," and to be selected on a balanced regional basis.[19] Its size was to be determined by each ordinary administrative conference, and its members were to serve in the interval between such conferences. The IFRB was envisioned as a kind of international court of justice for disputes concerning the use of the radio spectrum.

The concept was not new. As noted earlier in this chapter, an International Radio Technical Committee which would operate in a similar fashion was envisioned by the Inter-Allied Radiotelegraph Commission in 1919. The adamant opposition of the U.S. telecommunications industry coupled with the disappearance of the internationalists of the Wilson administration resulted in the demise of that plan. However, the atmosphere following the second World War which resulted in the creation of the United Nations and a host of other international organizations spilled over into the ITU arena. Thus, a concept which had withered away in the twenties was able to spring up again in the newly fertile ground of the forties.

The Atlantic City Telecommunication Conference produced more changes in the ITU structure than any conference in its history, past or present. Indeed, it established the modern ITU. In anticipation that the implementation of these changes would lead to a significant increase in the expenses of the Union, the conference decided to change the class-unit system of contribution to ITU expenses. A new thirty-unit class of payment was introduced, as well as a new one-unit class which allowed members to increase or decrease the amount of their contribution if they saw fit.

The major substantive changes which occurred at the Atlantic City conferences took place at the Radio and the High Frequency Conferences. Extensive changes were made to the major technical standard—the International Table of Frequency Allocations. These

---

[19]ITU, *International Radio Conference, Atlantic City, 1947,* Document No. 11 R, United States of America, Proposal for a Central Frequency Registration Board.

reflected the great advances in radio technology made during the war which allowed the upper reaches of the frequency spectrum to become usable, and made possible the extensive use of new services such as radar.

The more profound and controversial changes, however, had to do with the Union's rights vesting mechanisms, i.e., the way a nation secures the right to use a radio channel free from harmful interference by the stations of other nations. The subject of rights vesting will be discussed in greater detail in Chapter 12. The 1947 conferences saw an attempt made, principally by the United States, to completely revamp the scheme which had been employed for the previous forty years. That scheme had consisted of entrusting the Berne Bureau with the responsibility for maintaining a list of radio stations notified to it by member nations. Members were then obliged to operate stations so that they did not cause harmful interference to stations already on the Berne List. The presence of many non-existent stations on the list ultimately caused considerable concern, and led to United States initiatives at Atlantic City.[20]

The new scheme promoted by the United States called for the nations of the world to access their individual communications needs, and to work with U.S. experts to divide up the usable radio frequencies among the membership, based on sound engineering principles. Once accomplished, the new master plan would be entrusted to a Central Frequency Registration Board which would have the power to modify the list in accordance with the changing requirements of members. A separate and even more ambitious scheme was envisioned for high frequency broadcasting service involving a separate frequency plan and a board which would operate in conjunction with regularly scheduled planning conferences.

The other nations of the world, however, were not quite as willing to embark on this bold new adventure in international radio arrangements. What actually occurred at Atlantic City was the establishment of a Provisional Frequency Board (PFB) which would produce a new planned list during the months following the Radio

---

[20]See U.S., Department of State, Telecommunications Policy Staff, *Report of the United States Delegation to the Plenipotentiary Conference of the International Telecommunication Union, Atlantic City, 1947*. Serial No. 341, Washington, D.C., 1947 (mimeo.), p. 66.

Conference. That list would eventually be turned over to the permanent International Frequency Registration Board for maintenance. In the high frequency broadcasting area, the most that could be agreed upon was the creation of a small technical committee to prepare for another conference on the subject. In its final form, the IFRB enjoyed much less power than originally envisaged by the U.S.

After laboring for a long, hot summer, the delegates to the three Atlantic City conferences returned home, confident that many of the major technical problems confronting radiocommunication had been remedied, or that at least a good start had been made. The ITU's structure had been modified in such a manner that it would be able to cope with both the new telecommunication technology and the new international political climate in a fashion acceptable to the vast majority of the ITU members.

chapter 2

THE CHALLENGES OF CHANGE (1947-PRESENT)

The period of the modern ITU begins with the 1947 Atlantic City conferences. In that year, a coalescense of political power, technical developments, and a renewed belief in the utility of international organization brought the nations of the world together to produce new kinds of institutional arrangements for telecommunication. Although the character of the ITU was permanently altered, it took more than a decade of conferences to settle the changes set in motion in 1947. The first section of this chapter will deal with this post-Atantic City environment.

If the 1950s can be characterized by the turmoil involved in coping with Atlantic City developments, the 1960s can be regarded as a period of calm. The only significant new telecommunication development was the appearance of satellite radiocommunication which was at first only relevant to the major powers. Initial technical standards were established, but little else was done. This period is reviewed briefly in a second section of this chapter.

Although the sixties were relatively calm, indications in other forums such as UNESCO and the United Nations showed that the new developing nations which were joining the international community were interested in communication issues and that their perceptions often differed significantly from those of the older, industrialized nations. The 1947 changes in ITU membership requirements and the single vote per member rule, combined with the influx of scores of new developing nations, allowed a substantial shift in power and emphasis within the ITU. By the 1979 World

Administrative Radio Conference, there was no doubt that the
developing countries would try to make the Union more responsive
to their needs.

It would be wrong to give the impression that raw political power
was the only significant factor at work in reshaping the ITU.
Indeed, the on-going revolution in communications and information technology has had the most profound effect on the Union's
work. The technology, which is changing exponentially, continues
to produce pressures on the Union to keep technical standards relevant to the emerging transmission systems. Ultimately even the
basic methodology of establishing standards — dividing telecommunication into "services" and devising separate regulations for
each service —will be called into question by a number of technical
developments, including the integration of all services into a common global network. This is the subject of the last chapter.

The 1970s ended much as the 1940s had ended. A major conference
pointed the ITU in new directions. The stage was set for a series of
important conferences focussing on a wide range of telecommunication issues, and the future nature of many of the ITU's
institutional arrangements was unclear.

Calming the Waters (1948-1959)

The eleven year period immediately after Atlantic City was marked
by more conferences and meetings than had been seen in all eighty-
seven previous years combined. There were two plenipotentiary
conferences (1952 and 1959); thirteen regular and two special sessions of the Administrative Council; two administrative telegraph
and telephone conferences (1949 and 1958); two general administrative radio conferences (1951 and 1959); a two session administrative
aeronautical radio conference (1948 and 1949); two H.F. broadcasting planning conferences (1949 and 1951); three regional administrative radio conferences (1949); four CCIT Plenary Assemblies
(1948, 1948, 1953 and 1956 [the last]); four CCIF Plenary Assemblies
(1948, 1951, 1954 and 1956 [the last]); two CCITT Plenary Assemblies (1956 [the first] and 1958): five CCIR Plenary Assemblies (1948,
1951, 1953, 1956 and 1959); and scores of special committees and
study group meetings. All this activity can be subdivided into three
major categories: radio matters, telegraph and telephone matters,
and organizational matters.

Radio Matters

The principal radio activities during this period involved the struggle to implement the new arrangements introduced at Atlantic City for vesting rights to interference-free use of specific radio circuits. These new arrangements rested on the assembly of a new "planned" International Frequency List (IFL). Between the years 1948 and 1952, thousands of hours were spent at dozens of international conferences and meetings to draft a new IFL. The major forums for this effort were the Provisional Frequency Board (actually more of a conference than a board), the two high frequency broadcasting conferences, the aeronautical radio conference, the three regional radio conferences, and the Extraordinary Administrative Radio Conference. Despite the amazing effort—involving almost 80 months of conference work—a completely new list was never effected. Plans were agreed upon only in the aeronautical and maritime radio bands. It was left to the 1951 and 1959 administrative radio conferences to sort out the pieces and produce acceptable arrangements.

The task of producing the new International Frequency List and integrating the work of the other conferences was entrusted to the Provisional Frequency Board (PFB). The Board was established at Atlantic City to examine bands of frequencies between 10 kHz and 30 MHz which had been allocated by the Radio Conference to fixed, maritime, and aeronautical land services, and the assignment of specific frequencies to all the stations (by circuit) on the basis of the request lists submitted by all Union members. It met almost continuously during a 25-month period between January, 1948, and February, 1950, at Geneva, but failed in its task for the same reasons most of the planning efforts of this period failed: the submitted requirements greatly exceeded the available spectrum and the actual requirements were difficult to ascertain. In addition, there was the adamant opposition of the U.S.S.R. and its allies to the entire approach as an abridgment of their sovereignty.[1]

During much of the same period that the PFB was meeting in Geneva, separate specialized planning efforts occurred there and elsewhere around the world. One of the most arduous of these was

---

[1]See Statement by the Delegation of the U.S.S.R., *Extraordinary Administrative Radio Conference,* Geneva, 1951, Doc. No. 470.

the preparation of an allotment plan for shortwave broadcasting in which each country would get an equitable share of the appropriate frequencies, which were severely limited in number. The chief difficulty was that the concept of equitability was not the same for every nation.

The International High Frequency Broadcasting Conference at Atlantic City inaugurated the shortwave broadcasting planning effort. After that conference ended in late September, 1947, a Technical Planning Committee met in Geneva from March through June, 1948, to prepare for the second International High Frequency Broadcasting Conference at Mexico City. The Mexico City conference was a trial by endurance that continued under difficult conditions for six months between October, 1948, and April, 1949. The controversy centered on the General Principles Committee and the effort to reach agreement on the basis on which to allot channel-hours among all nations for their shortwave broadcasting activities. Although a plan was produced, a number of major powers, including the United States, refused to even sign the Final Acts, and the U.S.S.R. and its allies subsequently failed to participate in any further high frequency planning efforts. The Technical Planning Committee met again from June to December, 1949, and again briefly in March, 1950, just prior to the third International High Frequency Broadcasting Conference at Florence and Rapallo. The third conference met for nearly five months, only to agree that a plan could not be produced, and further work was needed.

The planning effort for aeronautical radio bands occurred at the Administrative Aeronautical Radio Conference. A preparatory committee gathered several weeks prior to the conference in late April, 1948. The first session of the conference met for four months from May to September 1948, followed by a second session which met for ten weeks from August to October, 1949. As has been characteristic in most aeronautical communication matters since the inception of those services, the relatively small community of experts involved, with the assistance of the International Civil Aviation Organization (ICAO), was able to accomplish its task without much rancor.

Administrative radio conferences were also held within each ITU region to supplement the planning efforts occurring at the PFB and the other international conferences.[2] The Region 1 conference met in Geneva for four months from May to September, 1949; the Region 2 conference met in Washington for three months from April to July, 1949; the Region 3 conference met in Geneva for six months from May to November, 1949. These multiple efforts did produce substantial agreement concerning plans for many radio services below 4 MHz and for the aeronautical and maritime services between 4 and 27 MHz.

All these efforts finally culminated in an Extraordinary Administrative Radio Conference which — after four postponements— was held in Geneva for four months from August to December, 1951. International tensions were at their peak. The Korean War was in progress, and the Cold War between the U.S. and the U.S.S.R. had begun. Originally scheduled to last three weeks, the conference went far beyond that period. After considerable vitreolic debate with obvious political overtones, it settled for the results achieved over the previous four years; it accepted the plans developed by the PFB for the fixed services between 14 and 150 kHz, those produced by the regional conferences for the bands below 4 MHz, and those developed by the Aeronautical Radio Conference. The EARC also completed the PFB's work on plans for the maritime mobile radiotelegraph and radiotelephone services between 4 and 27 MHz. For the future, it instructed the IFRB to develop draft plans for the high frequency broadcasting, fixed, land mobile, and tropical broadcasting services.

Thus, by the end of 1951, the concept of developing allotment plans had been retained, but the hoped-for new international frequency list consisted only of plan fragments. In the bulk of the bands, use continued on an unplanned basis.

After these decisions were ratified at the Plenipotentiary Conference at Buenos Aires the following year, the international radio regulatory scene remained painfully quiet until 1959. Only the IFRB, which continued work on the seemingly unachievable plans, engaged in any significant planning activity. In addition, a small regional conference was held at The Hague in 1957 to draw up an agreement

---

[2] The ITU's newly defined three regions for purposes of frequency management were: Europe, Russia, and Africa -Region 1; the Americas - Region 2; and Asia, Australia, and Oceania - Region 3.

concerning the use of short-range VHF communication between ships.

The next time the nations of the world assembled for a major radio conference was in 1959 at Geneva. The first few weeks of the general World Administrative Radio Conference of that year (WARC 59) went slowly, as much of the struggle following the Atlantic City conferences had to be recounted. At the outset, it was faced with the question: should the concept of a planned international frequency list be retained and should the conference attempt to finish the task which earlier conferences, the IFRB, and the PFB had failed to accomplish? A partial answer was available. The draft plans prepared by the IFRB for the high frequency broadcasting service were unacceptable to most countries, and the IFRB had been unable to compile a draft for fixed, land mobile, and tropical broadcasting services. In addition, it was recognized that what remained fell considerably short of being an accurate reflection of the actual use being made of the radio spectrum. It was apparent, therefore, that the conference could not successfully undertake the preparation of a complete plan in the time available to it, when previous attempts involving many years of effort had failed.

The conference confronted the problem in much the same way that the EARC had eight years earlier. It retained those portions of the list for those bands which had been planned and accepted by the EARC. For all the other bands, existing stations in use would be grandfathered. Administrations would be required to notify the IFRB of all new stations, the IFRB would subject them to a review procedure, then to enter them in the list. The only exception to this procedure was high frequency broadcasting which remained so contentious that the arrangements which were adopted involved little more than requiring nations to try to coordinate broadcasting schedules among themselves, with the administrative assistance of the ITU. Thus ended the longest and most difficult struggle in the history of the ITU, during which the idealism of the United States in planning the radio frequency spectrum finally succumbed to the pragmatism of world politics.

Although most of the radio regulatory activities of the 1948-1959 period involved the struggle over the Atlantic City rights vesting arrangements, other significant work occurred within the ITU also.

The CCIR continued to meet regularly under the 1947 Convention and became an increasingly important mechanism for developing international technical standards. Five CCIR plenary assemblies met during this period: the Fifth Plenary in Stockholm in 1948; Sixth Plenary in Geneva in 1951; Seventh Plenary in London in 1953; Eighth Plenary in Warsaw in 1956; and the Ninth Plenary in Los Angeles in 1959. During that time the number of Study Groups grew from eight to fourteen, and the standards and operating practices drawn up and adopted became more numerous and significant.

At the WARC 59, the assembled nations redid the most extensive of the radio technical standards, the Table of Frequency Allocations. Over the twelve years since Atlantic City, the state of the art had been constantly expanded to make ever higher radio frequencies available for use; the upper limits of the Table jumping from 10.5 GHz to 40 GHz. Also in 1959 a brand new radio service was defined that would eventually bring about a new era of international conferences and issues: the satellite radiocommunication service.

In summary, at the end of the twelve year period following Atlantic City, reasonably satisfactory global radio arrangements had been produced and the international telecommunication scene had become calm.

Telegraph and Telephone Matters

In contrast to issues relating to radio, those concerning telegraph and telephone matters during this period were subdued. Of course, there was no natural resource to be allotted among nations or other kinds of inherently contentious issues to be dealt with in the telegraph and telephone forums. They were concerned mainly with the technical standards and administrative procedures that were necessary to allow international telegrams and telephone calls to be placed.

This period saw the convening of two administrative telegraph and telephone conferences (Paris, 1948, and Geneva, 1958). No conference to address the Telegraph and Telephone Regulations had been held since the Cairo conference in 1938. A conference had been scheduled at Rome in 1942, but the war intervened. When the ITU members met at Atlantic City in 1947, they resolved to hold a Telegraph and Telephone Conference in 1949. The French government offered to host the conference, which was convened in Paris.

Both the Telegraph and Telephone Consultative Committees had met in Brussels the previous year to prepare for the Paris Conference. The CCIT created a special Committee of Eight to meet several months before the conference to draft proposals for amending the Telegraph Regulations so that they could be accepted by all members. Heretofore, the United States and a few other nations which had telegraph systems operated by private companies rather than the government had not signed the regulations because many of the provisions applied only to government operated systems.

Nearly seventy countries met for four months during the summer of 1949 at the Paris conference. After twelve plenary sessions and more than a hundred committee meetings, many significant changes were made to both the Telegraph and the Telephone Regulations. For the most part, these changes were very detailed, having to do with such things as letter codes, classes of telegrams, methods of accounting, and priority levels. In the end, enough changes were made to the Telegraph Regulations that, after nearly eighty-five years, the United States signed them for the first time.

Since the Telephone Regulations applied largely to the European system, including African and Asian nations which bordered Europe, some countries at the conference suggested that a second set of regulations be established to apply to intercontinental services. The United States, which had never operated under the Telephone Regulations, strongly objected to the proposal, supporting instead the efficacy of regional agreements. The matter was referred to the CCIF for study.

Nine years passed before another Telegraph and Telephone Administrative Conference was held. In the autumn of 1958, sixty-four ITU members met for three months to revise the Telegraph and Telephone Regulations. Although the conferences should have been held at five-year intervals, there had not been much enthusiasm for an administrative conference. The conference itself began to be referred to as the "status quo conference."[3]

---

[3] U.S., Department of State, Telecommunications Division, *Report of the Chairman of the United States Delegation to the Administrative Telegraph and Telephone Conference of the International Telecommunication Union, Geneva, Switzerland, September 29 through November 29, 1958*. TD Serial No. 893. Washington D.C., 1959, (mimeo), p. 10.

By the time it ended, however, the 1958 conference had made significant changes to the Telephone Regulations in order to give them universal applicability—a matter deferred at the 1949 conference. Committee 3 labored for many weeks at the conference to make the Regulations acceptable to as many countries as possible. This was largely achieved, with the notable exception that the United States still found them too restrictive and refused to sign the Final Acts.

The 1958 conference also raised a new issue. Several members felt that it would be useful to transfer most telegraph and telephone provisions from the Regulations to the CCITT Recommendations. As recommendations, the provisions would not have the force of a treaty, but in practice this meant very little. Since they were so widely followed and mandatory for international communication, the formalities of ratification as a treaty were superfluous. Moreover, as recommendations, the provisions could be under constant study in the CCITT study groups and amended at the much more frequent plenary assemblies. The idea was not adopted but was referred to the CCITT for further study. Fifteen years later, however, it was to be the major achievement of the 1973 Administrative Telegraph and Telephone Conference. This kind of slow evolutionary change over many years has been fairly characteristic of the ITU and the assoiated domestic regulatory bureaucracies which coalesce under its roof.

During the 1950s, the Telegraph and the Telephone Consultative Committees (CCIT and CCIF) held their Sixth through Eighth, and Fifteenth through Eighteenth Plenary Assemblies, respectively. The CCIT and CCIF held joint final plenary assemblies in November, 1956, closed the joint assemblies on 14 December, and reconvened the next day as a new, merged body, the CCITT. A special plenary was called during the 1958 Administrative Telegraph and Telephone Conference to consider the future work and organization of the CCITT.

All of these meetings were devoted to establishing the organization and administration of the consultative committees, preparing for administrative conferences, drafting regional network plans, studying the standardization of equipment, and adopting operating and maintenance procedures and technical specifications necessary for the proper operation and interconnectibility of telegraph and telephone systems throughout the world.

Organizational Matters

The 1948-1959 period resulted in a significant growth and matura-
tion of the ITU as an international organization. The changes
initiated at the Atlantic City Telecommunication (plenipotentiary)
Conference in 1947 were carried out and refined. By 1959, the structure of the Union and the nature of its activities were sufficiently
established to endure the next twenty years with little change.

The new International Telecommunication Convention drawn up at
Atlantic City invested the major power of the Union in the plenipotentiary conference, which was to meet every five years. In 1952, a
plenipotentiary was convened at Buenos Aires, but seven years
elapsed before the next one was held in Geneva in 1959. Under the
Convention, the Administrative Council operated as an executive
body to manage the affairs of the Union during the period between
plenipotentiary conferences. It normally met annually for approximately four weeks.

The Buenos Aires Plenipotentiary was convened in December, 1952.
Eighty-three of the eighty-nine members and associate members
attended the conference. After the many changes which had been
made at Atlantic City, there was little desire to take significant new
actions. In addition, the Cold War was at its peak and the international mood was not cordial. The major issues revolved around
attempts to plan the radio bands and the functioning of the IFRB.
The conference blessed the decisions made the year before at the
Extraordinary Administrative Radio Conference (EARC) which
had adopted the several plans which had been drafted and a notification and recordation scheme for the remaining radio bands. This
was a highly contentious issue, with the U.S.S.R unsuccessfully
urging that the IFRB be abolished and the Union revert to the 1939
Berne List of notified stations. The plenipotentiary also adopted
sanctions for those members which failed to become parties to the
convention within two years of its entry into force; changed the
character of the IFRB by providing for representation from eleven
countries rather than eleven independent individuals acting as
quasi-judicial officers, and by making the Board responsible for
additional duties to be assigned by administrative conferences or
the Administrative Council; froze the status of the Board by
transferring its terms of reference from the Radio Regulations to the
Convention; and adopted more detailed general regulations and

rules of procedure to govern the organization and conduct of conferences. Some controversy arose over the seating of delegates from the People's Republic of China and East Germany, and the matter was deferred to a future conference.

The next plenipotentiary was convened in Geneva in October, 1959 for nine weeks, concurrent with the meeting of the general World Administrative Radio Conference. Eighty-five of the ninety-six members participated.[4] Although most of the industrialized nations felt that few changes were necessary to the convention or the functioning of the Union, the new developing country members began to introduce changes to provide for their special needs.

It was at this conference that the Purposes of the Union were amended to include an entirely new function: to "foster the creation, development, and improvement of telecommunication equipment and networks in new and developing countries by every means at its disposal . . . " This new provision in Article 4 of the Convention was added at the insistence of the developing countries which, from this point onward, began to take an increasingly important part in shaping the role of the Union. Although technical assistance activities had been approved some years earlier by the Administrative Council, this amendment to the Convention represented a formal assumption of responsibility.

The size of the Administrative Council was increased from eighteen to twenty-five, based upon the increase in Union membership between 1947 and 1959 from seventy-six to one hundred and one. The regional distribution of the seats was figured on a mathematical basis, with the number of countries in each region determining the number of seats. The conference also decided that both the Secretary-General and the Deputy Secretary-General(the conference replaced the two Assistant Secretaries General with a single deputy position) should be elected by the plenipotentiary rather than by the Administrative Council.

A number of additional issues arose during the course of the conference. An attempt to relax the requirements of membership (U.N.

---

[4] Significant problems arose over the seating of delegations from China. Participation was denied to the People's Republic of China by a secret vote in plenary meetings.

membership plus two-thirds approval of members) was defeated.
The requirement that plenipotentiary and administrative conferences be held every five years was deleted, and the rules of procedure in
the Convention were made applicable to all Union conferences.

During the 1948-1959 period, the Administrative Council held fourteen regular and two special sessions. The special sessions were held
at the time of the plenipotentiary conferences to consider matters
relevant to those conferences and to inaugurate the new council.
Numerous issues important to the general operation of the Union
were considered by the council. The transfer of the General Secretariat from Berne to Geneva was approved for 1 January 1949; the free
distribution of documents was discontinued after 1 January 1949;
and the Provisional Frequency Board was terminated as of February, 1950.

At the sixth session of the Council in 1951, a resolution was adopted
regarding participation of the ITU in the United Nations Expanded
Program of Technical Assistance. In that resolution, the Council
confirmed a request made previously to the Economic and Social
Council of the U.N. for an allocation of funds for 1952; directed the
Secretary-General and the heads of the permanent organs to cooperate with the U.N. Technical Assistance Administration and
the Technical Assistance Board in responding to requests made of
the U.N. or the specialized agencies for technical assistance in the
field of telecommunication; and requested the permanent organs of
the ITU to explore the types and forms of techical assistance activity which might be undertaken in the field of telecommunication.
The action represented a landmark in the evolution of the Union. It
was embarking on an entirely new function — providing assistance
to developing countries. As the years passed and the composition of
the Union changed dramatically, this new function began to
assume growing significance.

In 1955 the Council approved the merger of the CCIT and CCIE, and
in 1956, it accepted the proposal of the Canton of Geneva to construct
a permanent home for the ITU (it had been temporarily located in
cramped quarters in the historic old Palais Wilson since the move to
Geneva). The foundation stone was laid during the 1958 Council
session. However, when the plenipotentiary met in late 1959, work
had still not proceeded beyond the foundation. After a minor furor

over this lack of activity by the Swiss and some dissatisfaction with the site, size, and design of the building, the Secretary-General was authorized to negotiate its purchase. At one point in the discussions, Mexico even offered Mexico City as an alternative site for the ITU![5]

A Placid Period (1960-1969)

The general character of the ITU during the 1960s was significantly different from that of the past. The tranquil atmosphere of the 1960s appears due to the retirement of many old hands from the international telecommunication scene, a feeling that twelve years of hectic activity in fashioning new international arrangements for the ITU were enough for a while, and a temporary lack of significant changes in telecommunication technology. While there were changes in some areas, such as the development of semiconductor devices and the emergence of satellite radiocommunication, they did not require extensive new international arrangements at that time.

Radio Matters

Five administrative radio conferences were held during this period: a 1963 world conference to consider satellite communication and radio astronomy provisions; a two-session world conference in 1964 and 1966 to revise the aeronautical frequency allotment plan originally drawn up in 1949; 1963, 1964, and 1966 African regional conferences to draw up frequency allotment plans for VHF/ FM and medium-frequency broadcasting; and a 1967 world conference to amend various technical standards pertaining to the maritime mobile service. All of these conferences brought together technical experts to achieve specific results, and were thus essentially free from controversy.

The Extraordinary Administrative Radio Conference to Allocate Frequency Bands for Space Radiocommunication Purposes was convened in Geneva in early October, 1963, for four weeks. In the few years since the 1959 general WARC, the use of artificial space satellites for radiocommunication began to grow, and international

---

[5]See U.S., Department of State, Telecommunications Division, *Report of the Chairman of the United States Delegation to the Plenipotentiary Confernece of the International Telecommunication Union, Geneva, Switzerland, October- December 1959.* TD Serial No. 905, Washington D.C., 1960 (mimeo), p. 27.

arrangements were necessary for their successful operation. In addition, radioastronomy had emerged as an important scientific discipline and required worldwide protection of certain frequencies from interference. The conference increased the allocations for space services from approximately one percent to fifteen percent and notification procedures were established to prevent harmful interference.

The Extraordinary Administrative Radio Conference for the Preparation of Revised Allotment Plan for the Aeronautical Mobile (R) Service was held in two sessions in Geneva. The first met for four weeks in January and February, 1964; the second met for six weeks in March and April, 1966. This two-session approach to the preparation of allotment plans was employed to allow the first session to concentrate on establishing the planning criteria and principles for allotting the channels. The second session was to adopt the actual plan based on these criteria and principles. The interval between the conferences also allowed the member administrations and the ITU staff to make the necessary preparations. Much statistical information concerning air navigation routes and operations was gathered to form the basis for the new plan. With no significant controversy, the conference adopted the new plan for high frequency radiocommunication along national, regional, and international civil air routes.[6]

The Aeronautical Conference also adopted a number of recommendations and resolutions concerning the use of VHF for aeronautical service and the broadcasting of meteorological information to aircraft in flight, the replacement of old double sideband equipment with single sideband gear, and it suggested the study of space radiocommunication for possible future use.

The African VHF/FM Broadcasting Conference met in Geneva in 1963 and produced a frequency plan for FM broadcasting stations in the African region. The following year, a similar conference was convened for medium wave (AM) broadcasting between 525 and 1605 kHz. Thirty-six delegations from African countries established the planning principles and technical criteria at this session. Adoption of the actual plan was deferred to a second session held in the

---

[6] See ITU, *Radio Regulations. Edition of 1976.* Geneva, 1976, Appendix 27. The new plan became effective in 1970.

autumn of 1966. The conference also examined the situation regarding low-frequency (150-285 kHz) broadcasting, but decided that no action was necessary at the time.

The last radio conference of the decade was the World Administrative Radio Conference to Deal With Matters Relating to the Maritime Mobile Service. It was convened in Geneva for six weeks in the autumn of 1967. The conference made substantial amendments to those parts of the 1959 Radio Regulations and the Additional Radio Regulations which applied to the maritime mobile service, including the gradual introduction of single sideband radiotelephony in many of the shortwave maritime bands, reduced channel spacing, and other specialized changes.

During this period, the Radio Consultative Committee (CCIR) held its Tenth Plenary Assembly in Geneva in 1963, and its Eleventh in Oslo in 1966. Through the work of its various study groups, scores of new studies were initiated and recommendations approved as the radio state of the art continued to rapidly evolve. It considered advances in space communications, produced an atlas of world ionospheric characteristics prepared by newly available electronic computers, recommended a VHF stereophonic broadcasting standard, and prepared a number of manuals to assist technicians in developing countries.

Telegraph and Telephone Matters

No administrative telegraph and telephone conferences were held during this period. The Telegraph and Telephone Consultative Committee (CCITT) held its Second Plenary Assembly at New Delhi in 1960, the Third at Geneva in 1964, and the Fourth at Mar del Plata, Argentina, in 1968. Increasingly, the work of the CCITT began to have worldwide applicability as more nations began to participate, and as its studies and recommendations achieved greater universality. Also during the course of these meetings, the concept rejected at the 1958 administrative conference concerning the transfer of most of the provisions of the Regulations to the CCITT's Recommendations began to gain favor.

Organizational Matters

During the 1960s only one plenipotentiary conference was held. The Administrative Council continued to meet annually to manage most of the organizational affairs of the Union. It should be noted, however, that this period witnessed a considerable increase in the number of members, all of them from the Third World. Although there were no significant controversies in the ITU, the stage was being set for a substantial shift in the balance of power — a shift that would manifest itself in the next decade in the form of attempts to make the ITU more responsive to the special needs of developing countries.

On May 17, 1965, the ITU marked its 100th anniversary with a special ceremony at the Salon de l'Horloge in Paris on the Quai d'Orsay where the original treaty establishing the International Telegraph Union had been signed in 1865.

Although no changes were made to the purposes of the Union, the development assistance activities received considerable emphasis throughout the conference. Dissatisfaction was voiced concerning the Secretary-General's past conduct of assistance activities, and Mohamed Mili from Tunisia was elected Secretary-General. Although proposals were made to establish regional ITU offices, a separate technical assistance fund, they were not adopted. Instead, Resolution Number 28 was passed which simply placed an additional emphasis on the importance of development assistance activities.

Between 1959 and 1965, the membership of the ITU increased from ninety-six to one hundred and twenty-nine. A number of delegations at the conference, noting the increase, proposed the addition of four more seats to the Administrative Council, three of which were to go to Africa and one to Asia. A secret vote resulted in an overwhelming approval of the changes.

Proposals were also made to integrate all of the headquarters staff under the Secretary-General and to abolish the IFRB. However, many nations feared the "one-man rule" which might result, and all such proposals were rejected in favor of less extreme solutions. It is noteworthy that the proposal to eliminate the Board, stridently advocated in the past by the U.S.S.R. was now introduced by the United States, which was responsible for its creation in 1947, as a quasi-judicial body with broad powers. Although the vote was sixty-four to thirty-nine in favor of retention, the move certainly did not enhance the prestige of the Board, and it was saved largely by the votes of developing countries. The size of the Board, however, was reduced from eleven to five members. The staff of the Union was not consolidated under the Secretary-General, but a Coordination Committee was formally established under his chairmanship, and he was given a more substantial hand in the activities of the specialized staffs of the CIs and the IFRB.

A proposal to convert the convention to a permanent constitution, and the plenipotentiary to a general assembly, failed. However, the matter was referred to a special group of experts for study and preparation of a draft constitution for consideration by the next plenipotentiary. The old scheme of ordinary and extraordinary administrative conferences was abolished; henceforth there would simply be administrative conferences. The conference was also plagued with the question of members' contributions in arrears. A plan was established to make partial but regular payments on the principal of the debt owed to the ITU. In addition, the contributory scheme was significantly amended by the adoption of a "one-half unit" contributory amount. This had the effect of allowing many developing countries to pay less by electing to pay a one-half unit rather than the previous minimun of a full unit and of requiring the industrialized countries to pay more if they retained large contributory units.

The accreditation of the Union of South Africa was an issue that gave rise to considerable political furor at the conference. Similar problems were not unknown in the past. For example, in 1927 the United States refused to allow the U.S.S.R. to attend the Washinton Radio-telegraph Conference because it did not like its form of government. This time, however, it was the Third World, and the preponderance of African and Arab countries in particular, which would not countenance the presence of South Africa at the conference. By a vote of fifty-seven to twenty-nine, the plenary voted them out of the conference and instructed the Secretary-General not to invite them to any future conferences.[7]

---

[7] The practice of excluding unpopular governments did not stop with the exclusion of South Africa. At the 1966 meeting of the Administrative Council, after telegraphic approval by a majority of ITU Members, Rhodesia was declared an illegal regime, and its signatures were deleted from the Convention.

During the sixties, the Administrative Council continued to meet annually, holding its 15th through 24th regular sessions. (In addition, a special session was held during the 1965 Plenipotentiary to inaugurate the new twenty-nine member council.) For the most part, the council's actions were fairly routine during this period, including devising conference schedules and agendas, setting budgets and staffing the secretariat. Some interesting highlights include the action by the Council in 1968 on the Montreux Plenipotentiary Recommendation to approve an extension to the ITU's existing building at Geneva. Its actions paved the way for the construcion of the new fifteen-story "ITU tower" overlooking Geneva and the lake.

Coping With a Changing World (1970 to Present)

The decade of the 1970s saw the Union in a new period of change, both technological and political, reminiscent of previous periods in the 1920s and 1940s. The difference, perhaps, was one of magnitude. Both the telecommunication technology and the composition of the Union changed so rapidly that some began to question the efficacy of the organization. The ITU did, however, begin to adapt to this new environment: a process which is still occurring.

Radio Matters

Between 1970 and 1981, there were seven administrative radio conferences: in 1971 to address world satellite communication matters; a two-session conference in 1974 and 1975 to prepare a low and medium frequency broadcasting plan for the European, African and Asian regions; in 1974 to revise the world maritime mobile plan and technical standards; in 1977 to prepare a world 12 GHz broadcasting satellite plan; in 1978 to revise the world aeronautical mobile plan; in 1979 to address most of the Radio Regulations in a general conference; and in 1980 to draw up the criteria and principles for a medium frequency broadcasting plan for the Americas which was adopted at a second session in late 1981.

The World Administrative Radio Conference for Space Telecommunications was held at Geneva for six weeks during the summer of 1971 to accomodate the rapid developments in the use of geostationary satellites for radiocommunication. This necessitated both changes to the ITU's technical standards, especially the Table of Frequency Allocations, and to the arrangements for using orbital positions and frequency channels without causing harmful interference. The latter was effected by the adoption of an advance notification and coordination procedure and would cause considerable concern for developing countries later in the decade.

One significant development was the identification of many different kinds of space telecommunication services, rather than a single basic space communication service as formerly, resulting in a significantly more complex Table of Frequency Allocations. The action occurred after extensive debate on its necessity, and may ultimately become anachronistic with the eventual implementation of dynamic digital satellite systems. A number of frequencies of special importance to radioastronomy were also recognized and offered protection.

Clearly the issues having to do with the acquisition of radio resource rights (also commonly referred to as orbit/spectrum rights), and the use of broadcasting-satellites (some of which can be designed to broadcast directly to special home receiving terminals, referred to as DBS), were the most significant and potentially contentious matters which arose at the conference. Although the industrialized countries' preference for a traditional notice and recording process was adopted,[8] the approach did not set well with many of the developing countries. To address their concerns, Resolution Spa 2-1 and Recommendation Spa 2-1 were adopted as a statement regarding equal rights of access to the resource, and a denial of any permanent acquisition of rights by those who would be the first to use the new rights vesting procedures. The recommendation further urged the next appropriate WARC to reexamine the whole process of rights vesting if countries began to "encounter undue difficulty" in gaining access to the resource.[9]

The use of direct television broadcasting by satellite had for some years been the subject of substantial debate in international forums outside the ITU, such as the U.N. and UNESCO.[10] These debates centered on the right of one nation to intentionally broadcast into another by satellite, without the recipient's permission (usually

---

[8] This process was, prior to the 1982 Radio Regulations, known as the Article 9A procedures, and will be discussed in greater detail in Chapter 11.

[9] This recommendation eventually set the stage for action at the 1979 WARC.

[10] See Kathryn M. Queeney, *Direct Broadcast Satellites and the United Nations* Leiden: Sijthoff Nordhoff, 1978.

referred to a "prior consent"). No resolution of the matter had been achieved in the other international forums. It mattered little that such a possibility was very unlikely due to the special receiving equipment which would be required by a large segment of the public. Many nations simply wanted international arrangements which precluded the possibility, for reasons of state sovereignty and cultural impact. Many socialist and Third World countries saw the adoption of a plan for the broadcasting-satellite service as an opportunity not only to obtain their "fair share" of the radio resource, but also as a means to circumvent the prior consent stalemate. The adoption of a plan for providing domestic DBS service would effectively preclude one country intentionally broadcasting into another. The European community also intended to implement a terrestrial microwave system at the same frequencies used for DBS, and wanted to plan the use of the spectrum to facilitate the design of that system. All of these interests resulted in the adoption of a resolution (Spa 2-2) calling for a future conference to plan the broadcasting-satellite service at 12 GHz, the most viable band for such a service. By a vote of one hundred to one the resolution was adopted. To complement this action a provision was also added to the Radio Regulations that required nations implementing a domestic system to use ". . . all technical means available . . . to reduce, to the maximum extent practicable, the radiation over the territory of other countries unless an agreement has been previously reached with such countries."[11]

In the autumn of 1974 and 1975, the Administrative Radio Conference for the Broadcasting Service Using Frequencies in the Medium Frequency Bands in Regions 1 and 3, and in the Low Frequency Bands in Region 1 was held at Geneva. The first session adopted the principles for planning, and the second drafted the actual plan. (During the interval between the sessions, the IFRB assisted in preparing a draft.) The new plan was comprehensive, replacing a host of old regional agreements, and establishing a channel spacing of 9 kHz for the world outside the Western Hemisphere. The only controversial issue involved was the reduction in number of low frequency broadcasting channels available to some European countries as a result of potential interference with African allotments.

---

[11] ITU, *Final Acts of the World Administrative Radio Conference, Geneva, 1979,* Geneva, 1981, The *Radio Regulations,* para. No. 2674. This provision was formerly known as No. 428A.

In the spring of 1974, the World Maritime Administrative Radio Conference met for six weeks in Geneva. The conference established a new allotment plan for high-frequency radiotelephone coast stations on the basis of single sideband operation. The old plan, prepared in the late forties, was very much out of date. The conference also adopted a large number of measures to enhance maritime mobile services, taking into account the trends toward an increase in radiotelephone traffic, a decline in manual radiotelegraph use, and an expansion in the use of radiotelegraph direct-printing systems.

The Maritime Conference was noteworthy for the increased involvement of developing countries, particularly toward assuring that the plan reflected their present and future needs, and the adverse reaction to this new involvement by Western industrialized countries, particularly the United States. The reaction was curious because the adopted plan on its face seemed equitable in meeting everyone's needs, and the U.S. had received more than 21 percent of the allotments, nearly four times more than any other country. In addition, the provisions accompanying the plan provided that if the allotments were not utilized within a fixed period, the nation holding an unused allotment would forfeit it. (This subsequently occurred.) Although four land-locked nations received allotments, all four obtained less than two percent of allotments, and three were industrialized countries with an arguable need for channels. The conference did seem to signal, however, that developing countries preferred to use planning conferences to vest rights in the radio resource, rather than to rely on the notice and recording mechanisms which still existed for many radio bands because of the failure of the Atlantic City and subsequent conferences to plan the entire spectrum.

In January 1977, a World Administrative Radio Conference for the Planning of the Broadcasting-Satellite Service in Frequency Bands 11.7-12.2 GHz (in Regions 2 & 3) and 11.7-12.5 GHz (in Region 1) was convened at Geneva for five weeks. The conference is generally known as the Broadcasting-Satellite WARC. Pursuant to the resolution adopted by the 1971 Space WARC, the conference proceeded to draft technical criteria, principles, and plans for the service. The conference rather amicably adopted a World Agreement and Associated Plan for Regions 1 and 3 (Europe, Africa, Asia and Oceania).

It also established provisions governing the service in Region 2 (the Americas) pending establishment of a plan, and recommended that a regional conference for preparing such a plan be held within five years.

The plan made allotments roughly proportional to the geographical area of countries.[12] Most nations of small to medium geographical size received four or five allotments. Those that were very large, such as China and the USSR, received allotments commensurate with their larger area. The notable exception to this action was the Americas. The United States convinced the small number of other countries present from that region to defer planning for several years, and to adopt interim procedures which included dividing the orbit into four alternating segments for the fixed and broadcasting-satellite services.[13]

The conference seemed to be a further example of the tendency of the developing country majority to use planning methods to obtain vesting rights. However, for this conference, the desire of the European nations to have a plan in order to proceed with terrestrial microwave systems, as well as the politics of international broadcasting of television without the consent of the recipient nation, were significant additional factors in producing the outcome.

In February, 1978, the World Administrative Radio Confence on the Aeronautical Mobile (R) Service convened at Geneva for four weeks. The conference produced an updated allotment plan for the aeronautical mobile (R) service based on new single sideband technology in the shortwave bands. No significant issues arose during the course of the conference.

The international radio event of the decade occurred in the autumn of 1979, when the general World Administrative Radio Conference (referred to as WARC 79) was convened at Geneva for ten weeks. The last such general conference had occurred in 1959; as a result, considerable domestic and international focus was placed upon WARC

---

[12] The allotments consisted of "nominal orbital position, channel number, boresight, antenna beamwidth, orientation of the ellipse, polarization, and e.i.r.p." See ITU, *Final Acts of the World Administrative Broadcasting - Satellite Administrative Radio Conference, Geneva, 1977.* Geneva, 1978 Art. 11. The allotment process and the definition of orbit/spectrum rights is discussed in greater detail in Chapter 11.

[13] This arc-segmentation approach was eliminated two years later at WARC 79.

79. More than two thousand delegates and supporting staff assembled to consider amendments to nearly the entire body of the Radio Regulations.

After a contentious beginning, which involved a one-week standoff over the chairmanship of the conference, a compromise candidate was accepted. This delayed the beginning of the conference four days and helped produce an atmosphere of lethargy which lasted several weeks. Ultimately, the pace of the conference picked up and significant work was accomplished. In this respect, however, the conference was similar to the general administrative radio conference in 1959, which went through a similar quiet period at the beginning when the labors of the previous twelve years to plan a new frequency list were recounted.

The most significant results of WARC 79 included the revision of many technical and operational standards for radio, particularly the Table of Frequency Allocations and the scheduling of a series of specialized conferences for the next decade. The table was expanded upwards from 275 GHz to 400 GHz, reflecting the advancing state of the art, as had been done at previous general radio conferences. It also modified the allocations in various frequency bands to reflect increased use of satellite radiocommunication. At the same time, however, many footnotes were introduced into the table to reflect different specific uses to be made by particular countries, the net effect of which was to decrease the global standardization of the use of the various bands and services. The long-term effects of rapidly-emerging technological developments which allow (if not require) substantial flexibility in the use of the radio spectrum, coupled with the desires of regional groups of countries to adopt their own special arrangements, were no doubt largely responsible for this trend. Despite attempts by one European delegate to discuss the impact of new technology on the efficacy of the allocation scheme, the subject was bypassed.

The import of WARC 79 was its display of the respective skills, needs, and power of developed versus developing nations, and the ability of the ITU to serve as a useful forum for negotiated international agreement. It also further confirmed the apparent intent of developing countries to make maximum use of planning mechanisms for vesting radio resource rights. This was evident prior to, during, and in the results of WARC 79, and was manifested in the resolutions to hold planning conferences for the shortwave broadcast bands and for the space radio services. In other bands where planning mechanisms were not sought, the developing countries demonstrated their ingenuity by drawing upon various preferential treatment norms adopted in many other international forums during the decade, and fashioning a scheme for being granted preferential administrative procedures for obtaining rights in the shortwave fixed service bands.

In the spring of 1980, the first session of the Regional Administrative Medium Frequency Broadcasting Conference (Region 2) was convened for three weeks in Buenos Aires. As is customary, its purpose was to establish technical criteria and planning principles to allow the IFRB to prepare a draft plan. It was followed by a second session in Rio de Janerio in November, 1981, to adjust and formally adopt the plan. It replaced a long-standing North American Regional Broadcasting Agreement plan. The only contentious issue involved was whether to adopt the 9 kHz channel spacing common to the rest of the world, or to remain with the existing 10 kHz spacing in the Americas. It opted for the latter.

Throughout this period, the Radio Consultative Committee (CCIR) continued to be very active, holding its Twelfth Plenary Assembly in 1970 at New Delhi, the Thirteenth at Geneva in 1974, the Fourteenth at Kyoto in 1978, and its Fifteenth at Geneva in 1982. Its activities encompassed the study of existing technologies, as well as emerging ones such as teletext and digital television.[14]

The Kyoto Plenary established a special interim working party to examine the future role of the CCIR, but the group does not appear to have wrestled with substantive issues or recommended significant changes.[15] These matters must remain for future plenaries to act upon. This response should be contrasted, however, with the apparent aggressivenes of the CCITT in meeting new issues and technology, as discussed below.

---

[14] ITU, CCIR, *Recommendations and Reports of the CCIR, 1978,*Geneva, 1979. ITU, CCIR, Decision 33 and Question 25-1/11.
[15] See *CCIR Report of CCIR Interim Working Party* PLEN/1, 1982.

Telegraph and Telephone Matters

One World Administrative Telegraph and Telephone Conference was held during this period—the first since 1959. It met in Geneva for one week in April 1973. The conference made substantial changes to both the Telegraph and the Telephone Regulations as recommended by the Fifth Plenary Assembly of the CCITT in 1972 and the Montreux Plenipotentiary in 1965. These changes involved the wholesale substituting of CCITT Recommendations for most of the provisions in the Telegraph and Telephone Regulations. This had been proposed at the last administrative conference in 1958 and, although the action was rejected at that time, it gained substantial support over the intervening years. The decision resulted in the United States signing the Telephone Regulations for the first time.

The CCITT held its Fifth Plenary Assembly in 1972, the Sixth in 1976, and the Seventh in 1980 — all at Geneva. The most dramatic changes occurred at the Seventh in the fall of 1980. Looking at the likely course of technological developments over the next decades, the CCITT completely revamped its study group structure and program around the Integrated Services Digital Network (ISDN) concept, with its Study Group 18 to play the lead role.[16] In addition, it took note of the desirability and importance of regional satellite systems and the non-technical needs of developing countries by the establishment of several new Special Autonomous Groups (GAS). The Seventh Plenary, recognizing the tremendous technological changes that will impact the world's "information transport" networks, boldly recommended that its jurisdiction be expanded to encompass all telecommunications except radio.

Organizational Matters

One plenipotentiary conference was held during the 1970s. It met for six weeks in the autumn of 1973 at Malaga-Torremolinos, Spain, to focus on the Convention and general affairs of the Union. One hundred thirty-one of the one hundred forty-six members attended. The results of the effort since the last plenipotentiary to produce a constitution were reviewed and largely rejected. A resolution was

---

[16] See, CCITT, *Questions Allocated to Study Group XVIII for the Period 1981-1984*, Doc. No. COM XVIII-No. 1 (February 1981).

adopted mid-way through the conference simply to divide the convention into two parts, the first to be known as the Basic Provisions grouping texts of a permanent character; and the second, the General Regulations grouping texts concerning the methods for conducting the affairs of the various organs.

The size of the Administrative Council was again increased — from twenty-nine to thirty-six — to reflect the growth in membership and the need for more representation from developing countries. The election of IFRB members was discussed, with the result that the next election was scheduled for the 1974 Maritime Conference. Thereafter, elections would be held at plenipotentiary conferences. The government of South Africa did not send a delegation to the 1973 Plenipotentiary because of its exclusion at the previous conference. The Plenipotentiary did, however, pass resolutions excluding both South Africa and Portugal, and eliminating Rhodesia's membership in the Union. The People's Republic of China was represented by a very large delegation, and was participating in an ITU plenipotentiary for the first time. Mohamed Mili, of Tunisia, was re-elected Secretary-General, and Richard A. Butler, of Australia, was elected Deputy.

As at Montreux, questions of technical cooperation received much attention. A special committee devoted its entire activities to this matter. Several resolutions were adopted which called for full participation in existing programs within the United Nations system. The issue of regional offices again engendered protracted debate, but, in the end, a resolution to create such offices was defeated. In the matter of funding technical cooperation activities, a special fund was established to be financed by contributions from member states. Also proposed was a new International Committee for Technical Cooperation in Telecommunications to coexist along with the CCIR and CCITT. The proposal was rejected in view of efforts in this area already being conducted within the existing Consultative Committees, especially the CCITT.

The provision of the Convention relating to the efficient and equitable use of the radio resource deserves special mention. The precise wording of this provision was the subject of extensive discussions. The industrialized nations placed emphasis on the phrase about countries having the "needs and the technical facilities at their disposal" as applied to use of the resource. The developing countries, on the other hand, placed emphasis on the phrase "equitable access."[17] Both phrases were added and in subsequent years, each group would point to this article to support their positions with respect to space services planning conferences.

During the 1970s, the Administrative Council continued to meet, holding its 25th through 35th regular sessions, and, as is customary, a special session was held at the 1973 Plenipotentiary to inaugurate the new Council of thirty-six members.

Conclusions

Although not for the first time, the evolution of the ITU stands today at a point of inflection. In the regulation of radio, three major cycles of change have occurred in the past sixty years alone. The exact nature of the changes that will occur in this latest cycle cannot be known at this time. What is known, however, is that nations must cooperate if communication is to take place. This fact makes the establishment of international telecommunication arrangements fairly unique in the field of foreign relations. The work of the ITU is more than merely desirable; it is mandatory. The continued existence of the ITU or its equivalent is no doubt assured as long as there are nation-states.

---

[17] ITU, *International Telecommunication Convention, Malaga - Torremolinos, 1973,* *Geneva, 1974.* Art. 33.

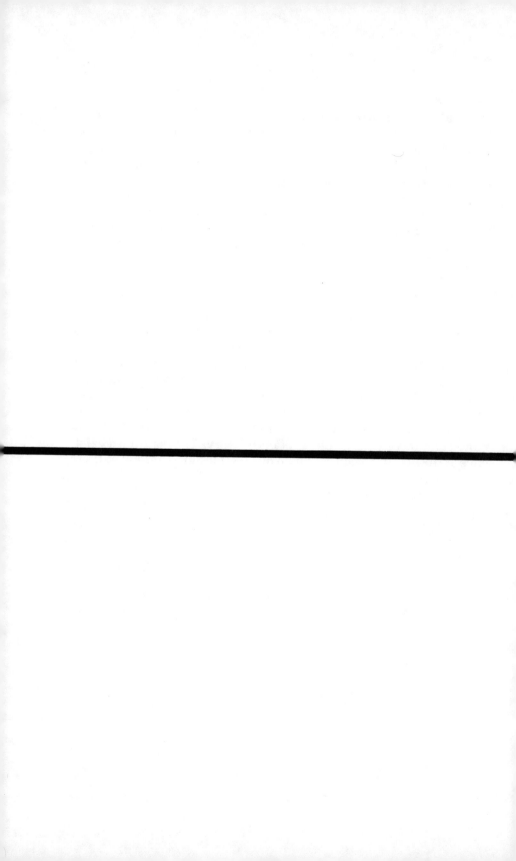

II

Decision-Making Structure

chapter 3

CONFERENCES

In the tradition of international organizations, to which the ITU was a major contributor, it is a fundamental rule that decision-making of importance is the prerogative of conferences of delegates from member countries. Governments tend to feel that they must have a voice, or the possibility of having a voice, in any decision that could affect them in an important way.

The ITU holds two major types of such conferences, the plenipotentiary conference and the administrative conference. The administrative conference category is further subdivided into world administrative conferences and regional administrative conferences. While the plenipotentiary has jurisidiction over all aspects of telecommunications, the administrative conferences customarily deal either with radiocommunication or telegraph and telephone matters.

We will begin with a close look at the plenipotentiary conference, including details of representation, structure, and rules of procedure. Since much of the procedure is the same for all conferences, the section on administratative conferences will concentrate only on those aspects which are different.

The Plenipotentiary Conference

The plenipotentiary conference is the supreme organ of the ITU, with ultimate responsibility, at least on paper, for the direction of the affairs of the ITU and international telecommunications.This authority stems directly from the fact that it is the plenipotentiary and only the plenipotentiary which has the power to revise the

ITU's basic treaty, the International Telecommunication Convention.[1] This instrument, as we shall see, gives the ITU its legal existence, establishes its structure and functions, and sets forth the basic regulations dealing with telecommunication in general and radio communication in particular.

In addition to the general task of revising the Telecommunication Convention, an instrument which will be discussed in some detail later, a number of tasks of a more specific nature have been given to the plenipotentiary to perform over the years.[2] Those of an administrative nature include the following:

1. To consider the report of the Administrative Council on the activities of the organs of the Union since the previous plenipotentiary conference;
2. To determine the fiscal limit for the expenditures of the Union until the next plenipotentiary conference;
3. To fix the basic salaries, salary scales, and the allowance and pension system for officials of the Union; and
4. To examine and approve (if necessary) the accounts of the Union.

The plenipotentiary conference is also directed to elect the members of the ITU to serve on the Administrative Council, to elect the members of the IRFB, ad to elect the Secretary-General and Deputy Secretary-General.

Five plenipotentiary conferences have been held since World War II: Atlantic City, July 2 to October 2, 1947; Buenos Aires, October 3 to December 22, 1952; Geneva, October 14 to December 21, 1959; Montreux, September 14 to November 12, 1965; and Malaga- Torremolinos, September 14 to October 25, 1973. The next plenipotentiary is scheduled for Nairobi in 1982.

Representation

Each country which is a member of the ITU has the legal right to be represented at a plenipotentiary conference. While most members

---

[1] The decisions of any plenipotentiary are authoritative in the sense that there is no body with the power to override them. However, decisions made by a plenipotentiary or any other representative body of the ITU or any international organization which affects the interest of a member, must be ratified by that country before it is binding on that country.

[2] The duties of the plenipotentiary conference are contained in the *International Telecommunication Convention, Malaga-Torremolinos, 1973*, Art. 6.

make a concerted effort to send representatives, there have been absentees at each plenipotentiary. For the four plenipotentiaries that have been held since the ITU was reorganized at Atlantic City, the attendance record has been the following:

Conference	members	attending
Atlantic City, 1947	78	76
Buenos Aires, 1952	87	82
Geneva, 1959	96	88
Montreux, 1965	129	122
Malaga-Torremolinos, 1973	146	131

In 1973 the absentees were: Colombia, Fiji, Guyana, Haiti, Honduras, Maldives, Malta, Nauru, Portugese Overseas Provinces, Qatar, Rhodesia, South Africa, Spanish Saharan Territory, Syria, and Tonga. (By resolutions passed at its 1967 and 1970 sessions the administrative Council had ruled that Rhodesia was no longer a member of the ITU.)

The South African situation was a special one. By 1965 the developing nations had achieved majority status in various international agencies and were beginning to make their presence felt. One of their primary targets of criticism was South Africa. This surfaced at the 1965 Montreux Plenipotentiary when the delegate of the United Arab Republic submitted a proposal on the part of a number of African delegates to exclude South Africa from the conference because of its racial policies. After a rather acrimonious debate, in which the delegate from the United States and others argued that the Telecommunication Convention did not permit the conference to exclude any member, the UAR proposal was passed and the delegation from South Africa withdrew from the conference.[3] A similar resolution was passed at the 1973 plenipotentiary, but as indicated above, South Africa, evidently in anticipation of such an action, failed to field a delegation[4]. Despite the violent criticism of these actions by the United States and some others, the fact that almost the entire membership of the ITU is represented at the plenipotentiary and that the delegates are equipped with plenipotentiary powers make the matter quite academic.

---

[3] See the account contained in Department of State, Office of Telecommunications, *Report of the United States Delegation to the Plenipotentiary Conference of the International Telecommunications Union, Montreux, Switzerland, September 14 to November 12, 1965,* T.D., Serial No. 973. Washington, D.C., 1965 (mimeo), pp. 9-11.

[4] For the 1974 resolution on the subject, see ITU, *International Telecommunications Convention, Malaga-Torremolinos, 1973,* Geneva, 1973, Resolution No. 31. A similar resolution was passed in 1973 excluding Portugese participation in ITU conferences; it was withdrawn after Portugal adopted a more democratic form of government. See *ibid,* Resolution No. 30.

While excluding a regular member from participation in plenipotentiaries and other conferences is almost unique, there have been a number of time consuming debates over the related issue of which government to invite in cases when two entities are competing for the title. In both the 1952 and 1959 Plenipotentiaries, it was the two China issue, and in the 1965 Plenipotentiary the U.S.S.R. argued for a "universalization" of the membership of the ITU, meaning the inviting of East Germany, North Korea, and North Vietnam.[5] Despite all of the assertions that as a technical organization, the ITU should be immune from "political" controversies, the history of the ITU since World War II shows that while political issues do not dominate the work of plenipotentiaries, the ITU is no more immune to politics than any other international organization.

Since the signing of the agreement in 1947 which made the ITU a United Nations Specialized Agency, a number of international organizations in the United Nations system have sent observers to ITU plenipotentiaries. UNESCO, which shares a major interest in communications with the ITU, has been present at all five post-World War II plenipotentiaries; the United Nations itself and the International Civil Aviation Organization have missed only one; the World Meteorological Organization and the Universal Postal Union have attended three; and the International Labor Organization and the World Health Organization have sent representatives to two ITU plenipotentiaries. The International Monetary Fund, the World Bank, the Intergovernmental Maritime Consultative Organization and the International ATomic Energy Agency have all attended once. A new body which made its appearance at Malaga-Torremolinos in 1973 was the United Nations Development Program.[6]

The participation of representatives from fellow international organizations is usually limited. The United Nations representative and the representative from UNESCO both usually give a short, formal speech dealing in generalities. They and the representatives of other agencies confine their other activities to informal discussions with members of the ITU Secretariat and delegations from member countries on items of mutual interest.

---

[5] For an account of the political problems at Atlantic City, see Codding, *The International Telecommunication Union*, pp. 208-223.
[6] Four international organizations were represented at the 1947 Plenipotentiary, five in 1952 and 1959, nine in 1965, and six in 1973.

It should be noted that the 1973 Plenipotentiary also passed a resolution permitting "liberation organizations recognized by the Uni-Telecommunication Union as observers."[7] No representative of such an organization presented himself at the Malaga-Torremolinso Plenipotentiary, but it did open the door for the first time to the direct participation in ITU plenipotentiaries of representatives of non-governmental entities.

Structure

The structure of an ITU plenipotentiary conference, or other ITU conferences for that matter, does not deviate greatly from that employed by other international organizations. The structure can be divided into three major components: the formal structure of the meeting itself which involves the delegates from member countries doing what delegates normally do, the informal structure which involves unofficial consultations between groups of delegates bound together by a particular political objective or other non-inclusive factor, and the structure represented by the Secretary-General and the members of his staff that are involved in the work of the conference. They will be treated in reverse order.

It would be virtually impossible to hold a plenipotentiary or any other large-scale conference without the intimate involvement of the Secretariat. It is the Secretariat that gathers the proposals that are submitted by the member countries and correlates them according to the parts of the Convention to which they refer; it produces the minutes of meetings and publishes the final report of the conference. In collaboration with the host government (when there is one), it arranges for meeting places, it provides for the necessary translation and interpretation services, it organizes the distribution of documents and gives expert advice to the chairman and vice chairmen of the conference, committees, and working groups.

The specifics of these tasks will be investigated in some detail in the pages which follow. At this point, however, it is appropriate to mention the lesser known and equally important role played by the Secretary-General in decision-making. The Secretary General produces documents both before and during the conference. He can

---

[7] See *International Telecommunication Convention, Malaga-Torremolinos, 1973,* Resolution No. 29.

emphasize certain of the agenda items and even downplay others, and he can produce documents supporting one course of action as against another. (The fact that most of the delegates are unfamiliar with conference procedures adds to the importance of such actions.) The Secretary-General also has control of the ITU's influential and widely read house publication, the *Telecommunication Journal.* The Secretary General also has an important role to play in the choice of the conference chairman and vice chairmen and the appropriate chairmen and vice chairmen of committees.

The second structure is made up of informal meetings of delegates to iron out a common position on subjects that come up in the course of a conference. This can include the ad hoc groups that may meet only once during a coffee break to discuss a particular subject that has arisen, as well as the more organized and long lasting groups such as the Non-Aligned Countries group, which can have a major influence on conference decision-making. Although a thorough knowledge of all such encounters would be beyond the capabilities of any investigator, there are some obvious ones that can readily be identified. At WARC 79, for example, meetings were held by several such groups —the Arab countries, African countries, equatorial countries, Caribbean and Latin American countries, ad the Non-Aligned group. One of the smallest was the group of equatorial countries which wished to push their claims for sovereignty over that portion of space above the countries used for geostationary communication satellites. The largest and best organized was the Non-Aligned Countries group, which met at frequent intervals and had a coordinating bureau of five delegates.

While groups of this type do not have a *de jure* sanction, they are becoming so common as to have a *de facto* sanction. The first plenary meeting of the 1973 Malaga-Torremolinos Plenipotentiary Conference heard an indignant statement from the Argentinian delegate complaining that he had sent a telex to the Secretary-tary-General schedule a meeting of the developing countries (the Group of 77) as early as possible during the Conference to enable them to adopt a common stand on "certain subjects of vital concern." "Unfortunately," complained the delegate from Argentina, "an unofficial document announcing such a meeting had been issued very late, with the result that it could not be convened before the meeting of the Heads of Delegations. It was to be hoped that such errors would not occur again."[8] It can be expected that as the size and complexity of ITU conferences continue to increase, such meetings will become more and more common.

The third structure is the formal one, made up of the conference plenary assembly and its chairman and vice chairmen, and the various standing committees headed by their own chairmen and vice chairmen. When there is a host government, the chairman of the conference is designated by the host government and is always an individual of some stature in that country's telecommunication administration. At the 1973 plenipotentiary that person was Leon Herrera Esteban, Spain's Director-General of Telecommunications. When there is no inviting government the chairman is selected by the heads of delegations who meet just prior to the opening of the conference to carry out this task.[9]

The conference vice chairmen, whose numbers have increased at each plenipotentiary, are selected by the heads of delegations on the basis of politics and geographical distribution. The 1973 Plenipotentiary chose the United States and Brazil from the Americas, Sweden from Western Europe, the German Democratic Republic and the U.S.S.R. from the communist bloc, Nigeria and Dahomey from Africa, and China and Japan from Asia and Oceania. The conference vice chairmen seldom have major responsibilities except for serving on the conference steering committee.

The major component of the formal organization is the standing committee structure and its leadership. The exact number of standing committees is fairly well dictated by the nature of the work of the plenipotentiary. All plenipotentiaries have a Steering Committee made up of the chairmen and vice chairmen of the standing committees and the chairman and vice chairmen of the conference, a Credentials Committee, and an Editorial Committee. Since 1952 each plenipotentiary has also had a conference budget control

---

[8] *Plenipotentiary Conference, Malaga-Torremolinos, 1973,* Document No. 94-E, 20 September 1973, p. 6. The Secretary-General replied that he would be happy to provide the necessary facilities for such a meeting but none of the Union's official documents required him to convene "unofficial meetings." The initiative should be taken by those who wish to be involved.

[9] There has been only one plenipotentiary without a host government since World War II; the 1959 Geneva Plenipotentiary. The heads of delegations chose J. D. H. van der Toorn of the Netherlands as the conference chairman.

committee. As regards the substantive side, in recent years it has been the practice to have committees devoted to Secretariat personnel questions, the ITU's finances, and either one or two committees dealing with the ITU's relations with the United Nations and development assistance. Plenipotentiaries also have at least two committees devoted to proposals for changes in the ITU's structure and matters concerning the general regulations for telecommunication and the more specific ones dealing with radio-communication contained in the ITU Convention.

Chairmen of the committees (with the exception of the Steering Committee which is chaired by the conference chairman) are selected by the Heads of Delegations meeting and based in about equal parts on finding competent committee leaders and providing good geographical distribution. With the exception of the Editorial Committee, which has always had a French chairman, no member countries seem to have had any special privileges. The Editorial Committee also has two vice chairmen, one from an English speaking nation, and one from a Spanish speaking nation, but no country other than Spain has been named to the vice chairmanship more than once.

The formal committee structure of the Malaga-Torremolinos Plenipotentiary was as follows:

Committee 1 — Steering
    Chairman and Vice Chairman of the Conference and Chairman and Vice Chairmen of Committees 2 to 9
Committee 2 — Credentials
    Chairman: Franisco F. Duarte (Paraguay), replaced on September 27 by Juan Balsevich (Paraguay)
    Vice Chairman: Emmanuel Egbe Tabi (Cameroon)
Committee 3 — Budget Control
    Chairman: M. K. Basu (India)
    Vice Chairman: Clinton A. Woodstock (Jamaica)
Committee 4 — Finances of the Union
    Chairman: Rudolf Rutschi (Switzerland)
    Vice Chairman: Ahmad Azheer (Pakistan)
Committee 5 — Staff Matters
    Chairman: F. Gerard Perrin (Canada)
    Vice Chairman: Ahmed Zaidan (Saudi Arabia)

Committee 6 — Technical Cooperation and Relations with the UN
  Chairman: Mohamed Benabdellah (Morocco)
  Vice Chairman: Ladislav Dvoracek (Czechoslovakia)
Committee 7 — Structure of the Union
  Chairman: Evan Sawkins (Australia)
  Vice Chairman: Laszlo Katona Kis (Hungary)
Committee 8 — Rights and Obligations
  Chairman: Gabriel Tedros (Ethiopia)
  Vice Chairman: Jose Hernandez-G. (Mexico)
Committee 9 — Editorial
  Chairman: Albert Chassignol (France)
  Vice Chairmen: Harold A. Daniels (United Kingdom)
         Jose Maria Arto Madrazo (Spain)

Seven high ranking members of the Secretariat were assigned by the Secretary-General to aid the committees in their work while the Secretary General acted as the secretary to the conference as a whole.

As we shall see numerous sub-committees ad working groups are usually created, although the ITU Convention admonishes conferences to form such groups "only when it is absolutely necessary."[10]

Rules of Procedure

There are a number of rules of procedure, other than those already mentioned, that should be reviewed at this juncture. While these rules do not differ greatly from those of other international organizations, there is enough that is unique to the ITU that a knowledge of them is important. As was the case in the preceding section, these rules pertain to administrative and plenipotentiary conferences.

Some rules apply to actions taken before the actual formal opening of the conference. Article 60 of the ITU Convention, for instance, requires that invitations be sent to all member countries by the inviting government one year prior to the opening date set by the Administrative Council. Replies regarding information on the composition of the delegations "must reach the inviting government not later than one month before the date of the opening of the

---

[10] *International Telecommunication Convention, Malaga-Torremolinos, 1973,* Art. 77, para. 4.

conference."[11] After the invitations have been sent, the Secretary-General requests that members send proposals for the work of the conference to him within a period of four months. The Secretary-General communicates all such proposals to all the other members of the Union as soon as they are received. The Secretary-General is also required to assemble and coordinate all the proposals that are received from members and all those from the plenary assemblies of the international consultative committees and to communicate the results to all members at least three months before the opening of the conference.[12] In order to be discussed in the conference, all such proposals and all proposals presented during the conference itself must be supported by at least one other delegation.

Although ITU conferences abound with statements to the effect that it is the tradition of the ITU to decide all matters of any importance by consensus, voting does occur and with some regularity. The rules are simple. All decisions except those concerning membership for countries not members of the United Nations are decided by simple majority vote. The majority is determined on the basis of those present and voting.[13] Abstentions are not taken into account when computing a majority , but for a valid vote to be taken in a plenary meeting "more than half of the delegations accredited to the conference and having the right to vote must be present or represented at the meeting."[14] Voting by secret ballot may occur when requested by five delegations and delegates may exercise one proxy vote.

Other rules of interest include the provision that members may submit "corrections they consider to be justified" to the minutes of plenary meetings (the only ones normally made available to the public) and to the summary records and reports of committees and subcommittees. In addition, the three official conference working

---

[11] If there is no host government, the Secretary-General is charged with convening and organizing the conference at the seat of the Union "after agreement with the Government of the Swiss Confederation." ITU, *International Telecommunication Convention, Malaga-Torremolinos, 1973*, Art. 64.

[12] The Secretary-General, the Directors of the Consultative Committees and the members of the IFRB do not have the right to submit proposals to conferences. See ITU, *International Telecommunications Convention, Malaga-Torremolinos, 1973*, Art. 66.

[13] Voting on new members is discussed on p. 128.

[14] *International Telecommunication Convention, Malaga-Torremolinos, 1973*, Art. 77, para 13.

languages are French, English, and Spanish. Other languages can be arranged for, but the expenses involved must be borne by those members requesting them.

Administrative Conferences

Although the ITU Convention provides for only two types of administrative conferences, world and regional, the structure and jurisdiction can vary considerably.[15] Some world administrative conferences have had the authority to revise most of the regulations, while others have treated only a small portion of the regulations. Three world administrative radio conferences having very broad jurisdiction have been held since World War II: in Atlantic City in 1974, in Geneva in 1959 and again in 1979. There have also been three world administrative telegraph and telephone conferences since World War II, the first in Paris in 1949, the second in Geneva in 1958, and the third also in Geneva in 1973.

In the same period it has been necessary to hold fourteen world administrative radio conferences with limited jurisdictions:

1. International High Frequency Broadcasting Conference, Atlantic City, 1947
2. International Administrative Aeronautical Radio Conference, 1st Session, Geneva, 1948
3. International High Frequency Broadcasting Conference, Mexico City, 1948-1949
4. International Administrative Aeronautical Radio Conference, 2nd Session, Geneva, 1949
5. International High Frequency Broadcasting Conference, Florence/Rapallo, 1950
6. Extraordinary Administrative Radio Conference, Geneva, 1951
7. Space Communication Conference, Geneva, 1963
8. Aeronautical Mobile (R) Conference, 1st Session, Geneva, 1964
9. Aeronautical Mobile (R) Conference, 2nd Session, Geneva, 1966
10. Maritime Mobile Conference, Geneva, 1967
11. Space Telecommunications Conference, Geneva, 1971

---

[15] See ITU, *International Telecommunication Convention, Malaga-Torremolinos, 1973*, Arts. 7 and 54.

12. Maritime Mobile Conference, Geneva, 1974
13. Broadcast Satellite Conference, Geneva, 1977
14. Aeronautical Mobile (R) Conference, Geneva, 1978

Several additional administrative radio conferences have been scheduled:

1. Two mobile services conferences in 1983 and 1988;
2. A conference for the planning of the high frequency broadcasting bands to be held in two sessions in 1984 and 1986:
3. A conference on the use of the geostationary satellite orbit and the planning of space services utilizing it, in two sessions in 1985 and 1987.[16]

A similar pattern is manifest in relation to regional administrative conferences. In the telegraph and telephone realm it has not been necessary to schedule any regional conferences since World War II. As regards radio, however, it was necessary to hold nine regional radio conferences between 1949 and 1979, and five more have been scheduled.[17]

World administrative conferences, general or restricted, can be convened by a plenipotentiary conference or a previous world administrative conference with the approval of the ITU's Administrative Council, at the request of one-quarter of the members of the Union, or on a proposal by the Administrative Council. With the exception of a decision by a plenipotentiary, all methods of calling administrative conferences require the concurrence of a majority of the members of the Union.[18] Regional administrative conferences can be convened by a plenipotentiary conference, on the recommendation of a previous world or regional administrative conference if approved by the Administrative Council, at the request of at least one-quarter of the members of the ITU belonging to that region, or

---

[16] See, *Final Acts of the World Administrative Radio Conference, Geneva, 1979,* Vol. II, Recommendation 12.

[17] For the purpose of radio frequency management, the world is divided into three general regions: Region 1 which includes Europe, Russia, ad Africa; Region 2 which includes the Americas; and Region 3 which encompasses Asia and Oceania. The ITU is currently considering the appropriateness of these regional divisions.

[18] *International Telecommunication Convention, Malaga-Torremolinos, 1973,* Art. 54. It should be noted that when the Administrative Council polls the members of the Union, those members who do not respond within the time limits set by the Administrative Council are not taken into account when computing the majority. See Art. 54. para. 6.

on a proposal by the Administrative Council. In the latter three cases the date and place must be ratified by a majority of the members of the Union from that region.

The agenda for administrative conferences is established by the Administrative Council with the agreement of a majority of the members of the ITU. (A majority of the members in the region for regional conferences.) The agenda must include any items that a plenipotentiary requests be put on it. The agenda or the date or place of an administrative conference may be changed upon the request of one-quarter of the members of the Union (one-quarter of the members of a region) or by a proposal of the Administrative Council if approved by a majority of the members (a majority of the region concerned). Once the agenda has been fixed, however, the ITU Convention specifies that the conference may discuss only those items. There is a serious controversy as concerns how restrictive this caveat is for a world administrative conference where all or almost all of the members of the ITU are represented by individuals with the power to speak for their governments.

Conference Procedures: WARC 79

One tends to think of an international conference as an isolated phenomenon, starting with the opening gavel and ending with the final thank-you speech. Most conferences, however are the culmination of years of prior effort and the work occasioned by the conference often goes on for another year after its formal closing. Further, because of the secrecy involved in ITU conferences and because the information that is given to the public is so restricted, it is seldom possible for the layman and perhaps even the government official who did not attend, to understand the true import of what transpired.[19] The work of the World Administrative Radio Conference at Geneva in 1979 will be used for illustrative purposes.

---

[19] For instance, only the official minutes of plenary meetings are ever made available for purchase by the general public. This represents only a fraction of the thousands of pages of documents issued during the course of a conference. The distribution of all other documents is restricted to the delegates. Further, it has been the tradition in plenary meetings that delegates may provide a written text of what they intended to say (sometimes not completely consistent with what was actually said) and those documents, and others which the public is not allowed to obtain, can be altered in case a delegate makes a mistake or says something he wishes he had not said. One delegate to a major ITU conference described these practices to one of the authors as those of "a mutual protection association," the implication being that delegates help each other to make their work at conferences as attractive as possible to their respective administrations.

The pre-conference activities for WARC 79 began in 1973 at the Malaga-Torremolinos Plenipotentiary when a working party under the chairmanship of a Norwegian delegate came to the conclusion that, because of rapid technical advances and changes in certain radio services and because of the necessity to harmonize the decisions of a number of conferences that had been held to deal with specific parts of Radio Regulations, it would be prudent to convene a conference to give the Radio Regulations a general revision. The plenipotentiary approved the working group's proposal and passed a resolution setting 1979 as the date and instructing the Administrative Council to make the neccessary preparations.[20]

From 1975 to 1979 parts of each meeting of the Administrative Council were devoted to WARC 79. In the 1975 session the Council directed that a circular letter be sent out to all administrations, asking for comments and suggestions concerning a possible agenda and the date and duration of the conference. At its 1976 meeting, with the help of the comments and suggestions that had been returned, the Council established an agenda that covered almost the entirety of the Radio Regulations, especially the revision of the important Frequency Allocation Table and most of the procedures for registering frequencies in the Master Register. The Council also decided that WARC 79 should convene in Geneva on September 24, 1979, for a period of ten weeks.

At both 1975 and 1976 meetings, the Council gave the CCIR and the IFRB directives for activities that would help prepare for the oncoming conference. In 1977 the Council went further and requested the CCIR to arrange for a special joint meeting of the CCIR Study Groups in October, 1979, to act as "a Special Preparatory Meeting (SPM) of the CCIR for the purpose of providing technical bases for the WARC 79 and for the use of administrations in preparing their proposals."[21] And, in October, 1977 the Secretary–General asked member administrations to inform him if they planned to send delegations to WARC 79 and, if so, how many delegates there would be.

By 1978 the preparations for WARC 79 were in full swing. At its 1978 spring meeting the Administrative Council established the budget

---

[20] *International Telecommunication Convention, Malaga-Torremolinos*, 1973, Resolution No. 28.

[21] ITU, Administrative Council, 32nd Session (May/June 1977), *Resolution 804*.

for the conference and in September the Secretary-General sent out invitations and requested that administrations return the proposals to him within four months so that they could be translated and circulated to other member administrations in time for them to be studied before the opening of the conference. Also in 1978 the CCIR held its special preparatory meeting. This meeting, which was held in Geneva from October 23 to November 17, 1978, was attended by 750 delegates and observers from 85 member countries, 30 recognized operating agencies, 18 international organizations, 10 scientific and industrial organizations, and three Specialized agencies of the United Nations. The work of the meeting was based on texts prepared by the XIVth CCIR Plenary Assembly and contributions from member administrations.[22]

In 1979 three regional preparatory meetings, called seminars, were held under the auspices of the ITU to discuss the results of the 1978 CCIR meeting and to present and discuss matters of regional interest. The first, for Arab and African countries, was held in Nairobi from February 12 to 23; the second, for Latin American and Caribbean countries, was held in Panama from March 12 to 23; and the third, for Asia and the Pacific, was held in Sydney from March 29 to April 10.

The list of preparatory conferences to WARC 79 would not be complete without a mention of the numerous pre-WARC 79 conferences that were held outside the auspices of the ITU, including meetings, sponsored by NATO, CEPT, CITEL, IMCO, ICAO, and the Telecommunication Coordinating Meeting of the Non-Aligned Countries for the Preparation of WARC 79 held in May, 1979, in Yaounde, Cameroon. The latter adopted a number of proposals that were considered beneficial to the lesser developed countries. An ad hoc group of individuals from eight of the non-aligned group was created at that meeting to meet on occasion to "bring out common or coordinated points of view."[23]

On the scheduled day, September 24, 1979, delegates from 140 member countries met in Geneva at the International Conference

---

[22] See ITU, *Report on the Activities of the International Telecommunication Union in 1978*, Geneva, 1979, p. 30.
[23] The group consisted of Kenya, Cameroon, Senegal, Algeria, Cuba, India, Iraq and Yugoslavia.

Center — just across the street from the ITU headquarters — to start work on some 14,000 proposals that had been submitted to amend the Radio Regulations.[24]

It was four days before the conference officially opened, however, due to an unprecedented struggle over the chairmanship. The normal unwritten procedure for the selection of a chairman for a conference without a host government, such as WARC 79, specifies that a series of unofficial consultations take place, usually orchestrated by the Secretary-General, to come up with a name that will be acceptable to the delegations which will be attending.[25] That name is then presented to the first heads of delegations meeting which convenes just prior to the inaugural meeting of the conference to prepare the first day's agenda. If that name is approved by the heads of delegations — which was the case almost without exception before WARC 79 — he is invariably accepted by the conference by acclamation.

In 1979 an event occurred outside the normal channels that would upset this routine. The meeting of the non-aligned nations had been held in Havana just a few weeks earlier had decided that the Chairman of WARC 79 should come from a non-aligned nation.[26]

For some reason this fact, and the fact that the non-aligned nations had a strong majority of votes at WARC 79, were overlooked and, at the Heads of Delegations meeting, the name of the head of the New Zealand delegation was presented by the Secretary-General. To the "surprise" and consternation of the Western delegates, the non-aligned nations countered with the name of T.V. Srirangan, head of the Indian delegation, and the only non-Western delegate who had previously

---

[24] Only fourteen members failed to send delegations: Bahamas, Bahrain, Barbados, Burma, Comoros, Equatorial Guinea, Laos, Mauritius, Sao Tome and Principe, South Africa, Surinam, Tonga, Trinidad and Tobago, and Viet Nam.

[25] If there had been an inviting government, that government would have named the Chairman. At Malaga-Torremolinos, 1973, the Spanish government named its Director-General of Telecommunications as chairman.

[26] See U.S. Department of State, Office of International Communications Policy, *Report of the Chairman of the United States Delegation to the World Administrative Radio Conference of the International Telecommunication Union, Geneva, Switzerland, September 24 - December 6, 1979,* TD Serial No. 116, Washington, D.C., 1980 (mimeo), p. 17.

chaired a world administrative radio conference.[27] Despite admonitions that ITU conferences had always chosen their chairmen by consensus, the non-aligned group continued to put forward Mr. Srirangan's name, and the Western bloc, under the leadership of the United States, countered with a series of other, mostly pro-Western, candidates. It appeared that if the non-aligned group would not go along with a pro-Western candidate "as a matter of principle," they would not go along with the first of the non-aligned group's nominees. Finally, late in the afternoon of Thursday, September 27, after a series of fruitless confrontations, the name of Roberto J.P. Severini, head of the Argentinian delegation, was suggested and accepted by both sides of the dispute. With the question of the chairman decided, the Heads of Delegations were able to confirm the agenda for the formal opening session and to compose a list of names for vice chairmen of the conference and chairmen and vice chairmen of committees to present to the conference.

WARC 79 was officially opened at the first plenary meeting on Thursday, September 27, 1979, by Per Mortensen (Norway), the dozen of the assembled delegates who had been chosen for this honor by the heads of delegations.[28] After a short speech from the acting chairman the group elected R.J.P. Severini as its permanent chairman by acclamation. After a short speech by the new chairman, the meeting approved without debate the election of six vice chairmen:

A. L. Badalov	(U.S.S.R.)
Jean Jipguep	(Cameroon)
Henry Kieffer	(Switzerland)
Linchuan Li	(China)
Angelo Petti	(Italy)
Glen O. Robinson	(U.S.A.)

---

[27] T.V. Srirangan had acted as Chairman for the 1978 World Administrative Radio Conference on the Aeronautical Mobile (R) service. The New Zealand delegate, Derek Rose, it should be noted, had been the Chairman of the 2nd Session of the Regional LF/MF Broadcasting Conference in 1975.

[28] The ITU Convention provides that in conferences without a host government the conference will be opened by the oldest head of a delegation. In the past the Secretary-General has actually polled each head of delegation to determine who was the oldest. At WARC 79 the Secretary-General of the ITU chose Mortensen on the basis of his appearance and former service in ITU conferences.

The Swiss vice chairman was also made the "coordinator" of the three principal technical committees. This action was followed by a lengthy speech by the Secretary-General and shorter ones by the Director of the CCIR and the Chairman of the IFRB.

The next step was the appointment of the committee chairman and vice chairman as follows:

Committee 1. *Steering Committee.* (Consists of the Chairman and Vice Chairmen of the Conference, and the Chairmen and Vice Chairmen of the other committees.)

Committee 2. *Credentials Committee.* Chairman: C. J. Martinez (Venezuela) and Vice Chairman: Amer Jomard (Iraq)

Committee 3. *Budget Control Committee.* Chairman: Z. Kupczyk (Poland), and Vice Chairman: L.P.R. Menon (Malaysia)

Committee 4. *Technical Regulations Committee.* Chairman: N. Morishima (Japan), and Vice Chairman: M. Cisse (Senegal)

Committee 5. *Frequency Allocations Committee.* Chairman: M. Harbi (Algeria), and Vice Chairman: J. J. Hernandez-G.(Mexico)

Committee 6. *Regulatory Procedures Committee.* Chairman: M. Joachim (Czechoslovakia), and Vice Chairman: E.J. Wilkinson (Australia)

Committee 7. *General Administrative Committee.* Chairman: P. O. Okundi (Kenya), and Vice Chairman H. L. Venhaus (F.R. Germany)

Committee 8. *Restructure of the Radio Regulations and the Additional Radio Regulations Committee.* Chairman: O. Lundberg (Sweden), and Vice Chairman: G. I. Warren (Canada)

Committee 9. *Editorial Committee.* Chairman: P. Bassole (France), and Vice Chairmen: V. Quintas (Spain) and D. E. Baptiste (United Kingdom).

After admitting observers from 34 international organizations which indicated an interest in participating in the work of the conference,[29] the conference decided on 9 a.m. to 12 noon and 2 to 5

---

[29] From the UN group there were representatives from the United Nations, UNESCO, World Health Organization, World Meterological Organization, and the Intergovernmental Maritime Consultative Committee. Three regional telecommunication organizations were represented, including the African Postal and Telecommunication Union (APTU), the Arab Telecommunication Union (ATU), and the Pan-African Telecommunication Union (PATU). Others included:

p.m. as the conference working hours and set Monday, November 19, at 10 o'clock as the date and time for the submission of the Credential's Committee's final report.

The day after the inaugural meeting the real work began. On a daily basis, the first order of business for most delegations was a meeting of all delegates to decide on apportionment of the work for the day, i.e., which meetings to attend and participate in. While the larger delegations had enough delegates to send more than one individual to each scheduled meeting, the smaller delegations, which were in the vast majority, had difficulty in covering even those meetings in which a subject was being discussed of special importance to its telecommunication administration and the problem was compounded as the conference progressed. The decision of the conference that meetings of more important committees would not be held at the same time was a help for the small delegations but, almost from the beginning, the major committees established sub-committees (working groups in the conference terminology) — four for Committee 4, six for Committee 5 and two for Committee 6. Further,

---

1. *Agence pour la Securite de la navigation aerienne en Afrique et Madagascar* (ASECNA)
2. European Space Agency (ESA)
3. North American National Broadcasters Association (NANBA)
4. International Air Transport Association (IATA)
5. Inter-American Association of Broadcasters (IAAB)
6. International Association of Lighthouse Authorities (IALA)
7. World Association for Christian Communication (WACC)
8. Intergovernmental Bureau for Informatics (IBI0
9. International Chamber of Shipping (ISC)
10. International Committee of the Red Cross (ICRC)
11. International Maritime Radio Association (CIRM)
12. International Special Committee on Radio Interference (CISPR)
13. Inter-Union Commission on Frequency Allocations for Radio Astronomy and Space Science (IUCAF)
14. International Electrotechnical Commission (IEC)
15. International Transport Workers Federation (ITF)
16. Arab Satellite Communications Organization (ARABSAT)
17. International Criminal Police Organization (INTERPOL)
18. International Radio and Television Organization (OIRT)
19. International Telecommunications Satellite Organization (INTELSAT)
20. International Space Telecommunication Organization (INTERSPUTNIK)
21. Asia-Pacific Broadcasting Union (ABU)
22. Arab States Broadcasting Union (ASBU)
23. Union of National Radio and Television Organization of Africa (URTNA)
24. European Broadcasting Union (EBU)
25. International Amateur Radio Union (IARU)
26. International Union of Radio Science (URSI)

not long after substantive work began, the committees resorted to a practice that has become more and more standard at ITU conferences, the farming out of anticipated or actual problem areas to even smaller sub-groups. Committee 5 alone created some twenty-three such groups. While this practice does get results, it also tends to further isolate the smaller delegations from the actual decision-making. (After the conference the Secretary-General estimated that during the ten weeks of the conference the number of plenary sessions and committee and major working group meetings totalled 894.)[30]

WARC 79 then settled down into a simple routine. The 14,000 proposals were apportioned out to the appropriate committees which were then assigned to the appropriate working groups. The working groups undertook their deliberations and, as problem areas were revealed or anticipated, sub-working groups were created as necessary to handle the issues involved.

As the sub-subworking groups completed their work, they presented their proposals to the working groups which then went into session to deliberate on the proposals and to consolidate the work of the various sub-subworking groups. Next the working groups made their reports to the full committees. For instance, on October 3, 1979, Working Group 5A made its first full report to Committee 5 and on October 4, Working Group 4A made its first full report to Committee 4. On October 20, 1979, the Editorial Committee (Committee 9) held its first full meeting to prepare to receive and edit the texts which would be submitted by the various committees to the plenary assembly. Of the three most important technical committees, Committee 4 made its first report to the plenary assembly on October 18, Committee 6 on October 19, and the hardworking Committee 5 on November 2.[31]

As soon as there were enough committee reports to warrant a meeting, the Steering Committee began to schedule meetings to the conference's plenary assembly to consider them. The first such meeting was held on November 8. Increasing numbers of plenaries

---

[30] See Mohamed Mili, "World Administrative Radio Conference," *Telecommunication Journal*, Vol. 47, No. 1 (January, 1980), p. 4.

[31] See ITU, WARC 79, Documents 179-E, 191-E, 223-E, 301-E, 305-E and 468-E.

were held on subsequent dates in November (13, 16, 22, and 24) until November 26 when plenary meetings were scheduled for each day until the final plenary which was held on December 5.

On December 6 a meeting of all the delegates was held to hear speeches from the Secretary-General, Jean Jipguep, Chairman of the Administrative Council and Vice Chairman of the Conference, and the Conference Chairman, R. J. P. Severini. Before the speeches, a formal ceremony was held for the signing of the Final Acts of the conference and the appended Final Protocol. One hundred and thirty-two delegations participated in the signing ceremonies.[32] Since the delegates had just finished approving reports of committees the day before, what the delegates signed was not the final version of the Radio Regulations which would come into force on January 1, 1982, but a hastily assembled Final Acts which would be put in its proper form and the signatures attached at a later date by the ITU's Secretariat.[33]

This action, however, did not complete the Secretariat's role, for it was still necessary to arrange for the final financial report and to continue to produce the documents. The last conference document, number 982, which contained the final list of participants, was not produced until February, 1980, and the final version of the Final Acts did not come off the press until 1981.[34]

Conclusions

The bulk of the decision-making in international telecommunication matters, at least until very recently, has been carried out by the traditional conference composed of delegations from member countries meeting at a pre-chosen location, often Geneva, with the help of the services of the ITU's Secretariat. It is here that the provisions

---

[32] Of the missing delegations, three never received their proper credentials, one had provisional credentials but never the final ones. Others were excluded for reasons such as being in arrears in making their contributions. See WARC 79, Docs. 730-E, 16 November 1979 and 981-E, 7 December 1979, p. 3.

[33] See, *Final Acts of the World Administrative Radio Conference, Geneva, 1979,* Vols. I and II, Geneva, 1979.

[34] For a comprehensive treatment of the work accomplished at WARC 79 from a United States perspective see U.S., Congress of the United States, Office of Technology Assessment, *Radiofrequency Use and Management: Impacts from the World Administrative Radio Conference of 1979,* Washington, D.C., 1982.

concerning telecommunications have been made, as well as those concerning the ITU's structure and functions. To become effective within the boundaries of member states, however, these decisions usually must be ratified by the competent authorities in each state.

This system, which is a common one in international organizations, is the product of tradition tempered by a serious respect for national sovereignty. The rules involved in such multilateral conferences are fairly simple and can be learned without much difficulty. Indeed, participation in one conference is often enough to make one an expert.

The material presented in this chapter highlights an emerging problem, however, one which could be especially serious for a standards-oriented international organization such as the ITU. As a result of the increase in the number of members of the international community, ITU conferences are becoming so big as to be unmanageable. Having a full-scale discussion of significant issues and obtaining a consensus is becoming increasingly difficult, especially on fundamental issues involving basic changes in the structure of the ITU that could allow it to better serve the needs of its clientele, or major changes in its functions to enable it to meet the challenge of exciting new technologies. Small-scale, partial remedies are substituted and then only with the increasing use, especially in radio conferences, of sub-committees, working groups, and sub-working groups.

In addition to the problem that this poses to the ability of the ITU to continue to adapt to changing conditions, it also tends to alienate the members who can only afford to field a small delegation and deprives those conferences of their potential contributions. This is not a healthy situation for an international organization such as the ITU, a remedy is needed.

chapter 4

CONSULTATIVE COMMITTEES

In order for governments and industry to take advantage of new telecommunication techniques and technologies, it is often necessary to agree to common standards. The study of innovations in telecommunication and the making of standards is the major activity of the ITU's International Radio Consultative Committee (CCIR) and the International Telegraph and Telephone Consultative Committee (CCITT).

These two bodies are composed of specialists from government and industry and carry out many of their functions through correspondence followed by relatively frequent meetings. Each has its own self-elected director and a specialized secretariat.

Background
Origins of the CCIF and CCITT

World War I produced a surge in interest in rapid and efficient communications that did not diminish at war's end. Telephone was one of the principal beneficiaries of this new interest. It became clear that there would be social, political, and economic advantages if the European countries could create an efficient telephone network linking capital cities and other areas of large populations. To make such a dream a reality, however, it was necessary to adopt common standards.

One of the early advocates of the need for an organ to bring the European telephone administrations together to set standards was Frank Gill, President of the British Institute of Electrical Engineers. It was Paul Laffont, Under-Secretary of State for French Posts and Telegraphs, however, who was responsible for the convening of a "Preparatory Technical Committee on International Telephony in

Europe" which met in Paris in 1923 with representatives of six European telephone administrations in attendance.[1] The agenda of this meeting contained four major items:

1. Selection of desirable characteristics for long-distance international telephone lines;
2. Determination of operating and maintenance methods for long-distance international telephone lines;
3. Drawing up of a program of work for establishing an international telephone network capable of meeting European needs;
4. Constitution of a central controlling organization for the development of European international telephony.

After considerable discussion, it was agreed to issue a recommendation that the Preparatory Committee become a permanent body responsible for preparing the organization and development of a European telephone network. It also recommended that no attempt be made to create a single European telephone administration for that network, but to keep both administrative and financial control in the hands of individual national administrations.

At the first meeting in 1924, it adopted the name International Consultative Committee on Long-Distance Telephone Communications,[2] decided to meet in principle once a year, and decided to create a permanent secretariat to be based in Paris. It also agreed to create a standing committee composed of delegates from the interested countries to consider projects for new international telephone links submitted by administrations.[3]

By the time of the International Telegraph Union's Paris conference in 1925, the Telephone Consultative Committee had more than proven its value in the establishment of standards for underground

---

[1] See Robert Chapuis, "The CCIF and Development of International Telephony (1923-1956)," *Telecommunication Journal,* Vol. 43, No. 4 (March, 1976), p. 186. In his article Chapuis also notes that individuals from various European administrations had met in Budapest in 1908 and Paris in 1910 as the First and Second Conference of Telegraph and Telephone Engineers to discuss common problems..

[2] In effect, the group adopted the French name *Comite consultatif international des communications telephoniques a grande distance* which was later abbreviated to *Comite consultatif international telephoniques.* The acronym CCIF is derived from the French name.

[3] At its first meeting the Standing Committee divided its work into: transmission; line maintenance and supervision; traffic; and operations. Chapuis, *op. cit. p. 187.*

telephone cables in the European network. Without any difficulty, that conference decided to bring the committee into a relationship with the International Telegraph Union. In consideration of the feelings of the telegraph partners in the European telegraph and telephone administrations, the Paris Conference also decided to create an International Telegraph Consultative Committee. While the actions of the Paris Telegraph Conference provided the two committees with the imprimatur of the International Telegraph Union, in effect they remained almost completely independent until 1947. About the only connection that existed between the two in the 1920s was two provisions: that any administration whose country was a member of the International Telegraph Union had the right to participate in their work and that the two committees were to communicate their recommendations to the Bureau of the International Telegraph Union in Berne which was given the task of publishing them.[4]

Over the years the two consultative committees adopted a similar structure. The major organ was the plenary assembly, to which all members were parties, which met in principle once a year. Each committee also had a series of working groups (called Committees of Rapporteurs) which brought together experts to study specific problems, mainly by correspondence, and make their recommendations to the annual plenary assemblies for their adoption or rejection. The expenses of the committees were met by the administrations participating in their work. The basic difference between the two was that the Telegraph Committee used the Telegraph Union's Bureau for support purposes while the Telephone Committee had an elected director, its own secretariat and, after 1928, its own laboratory.

Origins of the CCIR

The origins of the International Radio Consultative Committee go back to the First World War. The rapid advances in technology, the demonstrated value of radio, and the success during the war of the cooperation between the victorious powers in dealing with radiocommunication problems; all pointed to the need for a formal, continuing intergovermental body to study and make recommendations on international radio matters.

---

[4] See Codding, *The International Telecommunication Union*, pp. 47-48.

The first manifestation of such a body was in the form of a proposal submitted to a meeting of Allied radio experts to form an International Technical Committee "charged with giving advice on questions relative to the radiotelegraph service."[5] In subsequent meetings the name was changed to the International Technical Radiotelegraph and Visual Committee (C.I.R.V.) and it was suggested that it should become a distinct organ within a proposed overall Universal Electrical Communications Union.[6] However, a sudden change in attitude toward international cooperation due to a change in the U.S. Government resulted in 1922 in the demise of the entire undertaking.

An international radio consultative committee was finally created by the Washington Conference of the International Radiotelegraph Union in 1927. By that time, an additional reason had arisen for creating such a body. Because of the rapidly expanding nature of radio technology, there were always problems that had to be worked out at each radio conference before decisions could be made. "One of the main arguments for setting up the *Comite consultatif international technique des communications radioelectriques* was 'to undertake studies and present conclusions to the next conference, thus eliminating part of the burden of exhaustive technical studies that had been necessary during conference time.' Equally important was revising the Frequency Allocation Table."[7]

However, there was a great deal of opposition to the creation of the CCIR.[8] The U.S. delegate presented the argument that since radio technology was evolving so rapidly, its progress might be held back if this committee should establish too rigid standards. The French were afraid that private companies would obtain the approval of the committee and use it for commercial gain. The British registered

---

[5] *EU-F-GB-I Final Protocol of 25 August 1919*, p. 3. The Committee was specifically charged with giving "its opinion relative to the distribution of wave lengths used between large stations. It shall also examine the propositions relative to modifications to be made in the Convention, in order to prepare the work for the next International Conference. The Technical Committee shall have the right to decide questions relating to the abbreviations annexed to the Service Regulations." In subsequent meetings, through 1920, still more powers were added.

[6] See Universal Electrical Communications Union, *Draft of Convention and Regulations*, Art. 19.

[7] *From Semaphore to Satellite*, p. 156.

[8] Later to be reduced to *Comite consultatif international des radiocommunications;* hence, its acronym CCIR which is still used.

disapproval on the grounds that since no changes could be made to the radio regulations between conferences, the committee would be useless.

The French were afraid that private companies would obtain the approval of the committee and use it for commercial gain. The British registered disapproval on the grounds that since no changes could be made to the radio regulations between conferences, the committee would be useless.

The decision to establish the CCIR narrowly won approval at the Sixth Plenary Assembly of the 1927 Washington Radiotelegraph Conference. A general article was added to the Radiotelegraph Convention establishing a committee "for the purpose of studying technical and related questions" submitted to it by the participating administrations or private enterprises. Participants were responsible for contributing equal shares to the expense of the committee and the committee was to meet, in principle, every two years. All other details of its organization were left to the committee itself.[9]

The proposal won by a vote of thirty to twenty-six. It is interesting to note that the Germans were able to deliver all five of their colonial votes in favor of the proposal while Great Britain's vote was split when New Zealand and British India voted in favor.

The CCIR quickly adopted an organization similar to that of the telegraph and telephone committees, with a plenary assembly, working groups, and an elected director. The First Plenary Assembly met at The Hague in 1929 to discuss ten problems for study: definitions of power and frequency ranges; frequency measurements; monitoring band tolerance; occupied bandwidth; frequency separation; spurious emissions; fading; directivity; atmospheric noise; and amateur licenses.

The CCIs After 1927

Very few changes occurred in the organization or operation of the CCIs until the Atlantic City Plenipotentiary in 1947. The Madrid conferences of the Radiotelegraph Union and Telegraph Union, which created the International Telegraph Union, recognized the existence of the CCIs but left their organizations to the separate

[9] Codding, *The International Telecommunication Union*, p. 122.

[10] Article 16 of the Madrid Convention simply stated: "Consulting committees may be formed for the purpose of studying questions relating to the telecommunication services."

regulations and the conferences that made them.[10] The Cairo Administrative Radio Conference of 1938 discussed and rejected a suggestion that the CCIR be given the power to involve itself directly in rate questions; while the Cairo Administrative Telegraph and Telephone Conference charged the CCIT "with the study of rate questions submitted to it by a plenipotentiary or administrative conference or by at least twelve participating administrations."[11]

The Atlantic City Telecommunication Conference of 1947 spent much of its time considering the proper place of the CCIs in the ITU's organizational structure. On the one side were those who felt that the telecommunications of the future demanded that the consultative committees be brought into closer relationship to the ITU and that their organization and operation be standardized. Most of the delegations were in agreement, but many worried that in so doing they should not deprive them of the characteristics, including independence, that permitted them to work effectively. According to the Swiss delegate, the oldest of the committees, the CCIF, had been able to do so well over the years because it had been allowed to work "with complete independence and to adapt itself, according to the various cases, to the continually changing needs created by the inevitable evolution of engineering techniques."[12] The Atlantic City plenipotentiary first brought the CCIs into permanent relationship with the ITU by making the three committees permanent organs and making Geneva their permanent seats. The Atlantic City conference set one criteria for membership for all three committees: administrations of members and associate members of the Union and "recognized private operating agencies which express a desire to have their experts participate in the work of these Committees."[13]

It was affirmed that all three committees should work through a structure of a plenary assembly and study groups with a director appointed by the appropriate plenary assembly. The Telegraph and Radio Committees were given secretariats and elected directors to make them similar to the Telephone Committee. Rules of procedure were also established for all of the committees but permission was

---

[11] See Codding, *The International Telecommunication Union*, p. 175.

[12] As quoted in Codding, *The International Telecommunication Union*, pp. 297-298.

[13] *International Telecommunication Convention, Atlantic City, 1947*, Atlantic City, 1947, Art. 8, para. 3.

given to adopt any additional procedures that they deemed necessary. All three committees were given the task of studying technical and operating questions, and the Telephone and Telegraph
Committees were given the additional task of studying "tariff
questions."

The only major change in the CCIs that has occurred since Atlantic
City was the merger in 1956 of the International Consultative
Committee on Telegraph and the International Consultative Com-
mittee on Telephone to form the present International Consultative
Committee on Telegraph and Telephone (CCITT).

The proper number of consultative committees has long been a
subject of controversy in ITU circles. Italy, for instance, submitted a
far reaching — although unsuccessful — proposal to the Madrid
conference of 1932 which envisaged the establishment of a single
committee divided into three sections, one for telephone, one for
telegraph, and one for radiotelegraphy.[14] The Atlantic City Telecommunications Conference had debated a Norwegian proposal to
combine the CCIT and the CCIF on the basis that the administration of the two services were in the same governmental body in
most countries and that the technology had begun to merge. The
Atlantic City Conference also debated a U.K. proposal to create a
new International Broadcasting Consultative Committee, in view of
the rapid increase in broadcasting services and its unique problems,
and, more importantly, because the U.I.R., which had been performing consultative committee-related work for broadcasting, was
being disbanded. Although the first proposal was rejected, it was
decided to direct the two committees to study the question and
submit a report thereon to the next plenipotentiary. The result of
this report was the decision of the Buenos Aires Plenipotentiary to
merge the CCIT and the CCIF; this was carried out in 1956. As
regards the U.K. proposal, in view of the sentiment that such an
action would unnecessarily increase the complexity of the ITU as
well as its expenses, it was decided only that the CCIR should have a
permanent vice director who would be a specialist in broadcasting.
This requirement was abolished by the Geneva Plenipotentiary of
1959 but the CCIR did create a permanant study group dedicated to
the study of broadcasting related problems.

---

[14] See ITU, *Documents of the Madrid International Telegraph Conference (1932)*,
Berne, 1933, Vol. I, pp. 836-837.

Two proposals were made in the 1960s for a total restructuring of the consultative committees. In 1963, Secretary-General Gerald C. Gross suggested a plan for the reorganization of the Union. The CCIs were to become the Department of Research in the Secretariat. Although the work of the CCIs was to continue as it was, the CCIR and CCITT study groups would no longer exist. Instead there would be ITU research groups. This would eliminate the need for the plenary assemblies, as the results of the work of the Research Groups would be transmitted directly to conferences for approval.[15]

In 1964, John Gayer, former Chairman of the IFRB, put forth a plan to integrate the headquarters of the ITU by making the directors of the CCIR and the CCITT directly responsible to the Secretary-General. This action, he argued, would allow for maximum efficiency and effectiveness in the ITU as a functional organization.[16]

More recently, expressions of dissatisfaction have been leveled against the consultative committees regarding their structure and decision-making process: (1) The process of formulating recommendations is too time-consuming in view of the rapid pace of technological advance in telecommunications. The critics claim that the process of mutual adjustment in the study groups is slow and that the time lag between plenary assemblies can delay the production of equipment or result in the production of incompatible equipment.[17] (2) Often recommendations (mainly those of a technical nature) simply "register" the *de facto* state of affairs in industrialized countries. The recommendation thus becomes a hybrid monster as administrations attempt to include their national specifications. The resulting recommendation thus fails to fulfill the standardization function which is a primary aim of the consultative committees.[18] And (3) the consultative committees pay insufficient atten-

---

[15] Gerald C. Gross, "The New ITU," *Telecommunication Journal,* Vol. 29, No. 10 (October, 1963), p. 311.

[16] John Gayer, "Past and Future — Integrating the ITU Headquarters," *Telecommunication Journal,* Vol. 30, No. 6 (June, 1964), p. 164.

[17] Harold K. Jacobson, "The International Telecommunication Union: ITU's Structure and Functions," *Global Communications in the Space Age: Towards a New ITU,* (New York: The John and Mary Markle Foundation and the Twentieth Century Fund, 1972,) p. 47.

[18] Raymond Croze, "Former CCITT Chief Answers IPTC Newsletter," *IPTC Newsletter No. 36* (February, 1977), p. 2.

tion to the problems and interests of the newly emerging countries. Because the participants generally come from the technically advanced nations, their recommendations are the result of an adjustment process in which economic and technical trade offs are made to suit their developed status.

Finally, the Argentinian delegation at the 1973 Malaga-Torremolinos Plenipotentiary offered a proposal on the floor to create a consultative committee devoted exclusively to problems of technical assistance. Although there was some support for the idea, as there has been for others, it was felt that there was too little time available to give it the consideration it deserved.[19]

Structure and Functions

The functions of the consultative committees and their basic structures are set forth in the International Telecommunication Convention. Article 11 of the 1973 Convention states simply: "The duties of the International Radio Consultative Committee (C.C.I.R.) shall be to study technical and operating questions relating specifically to radiocommunications and to issue recommendations on them." Article 11, paragraph 1 (2), states: "The duties of the International Telegraph and Telephone Consultative Committee (C.C.I.T.T.) shall be to study technical, operating *and tariff questions* relating to telegraphy and telephony and to issue recommendations on them."[20]

Questions studied by the committees originate in the plenary assemblies or are referred to them by a plenipoteniary conference, an administrative conference, the Administrative Council, the other consultative committee, and the IFRB.[21] The CCIs may study and offer advice on telecommunication problems of a single nation if that nation should so request. The only directive specifying subjects that should be studied is the admonition that the CCIs shall "pay due attention" to questions "directly connected with the establish-

---

[19] For the proposal see ITU, Plenipotentiary Conference, Malaga-Torremolinos, 1973, Doc. No. 96-E, 20 September 1973.

[20] *International Telecommunication Convention, Malaga-Torremolinos,* 1973, Geneva, 1974. Italics supplied.

[21] Questions may also be submitted by members of the ITU between Plenary Assemblies if at least twenty other members agree.

ment, development, and improvement of telecommunication in developing countries in both the regional and international fields."[22]

The structure of the consultative committees has not changed greatly from what it was in the beginning. Each of the committees has a plenary assembly which is its highest organ and which meets, according to the Convention, every three years.[23] Because of the importance of the work of the committees to administrative conferences, the convention provides that plenary assemblies should be convened, if possible, at least eight months before such conferences.[24] The detailed work is carried out by an elaborate infrastructure composed of study groups, committees, and working parties, the exact number of which is determined by the plenary assembly. In 1980, at the uppermost tier in this structure, the CCITT had fifteen study groups, one Joint CCITT/CCIR study group, five plan committees, and six special autonomous working parties; the CCIR had eleven study groups and two Joint CCIR/CCITT study groups. The plenary assembly also appoints the chairman and vice chairman of each group.

Both committees make use of working parties, in effect sub-working groups for the various study groups. The convention permits the committees to form joint study groups. The convention also specifies that the two committees form a joint World Planning Committee and such regional planning committees as are deemed necessary to "develop a General Plan for the international telecommunication network to facilitate coordinated development of international telecommunication services."[25] The study groups conduct their work as far as possible by correspondence but the convention does provide that they may hold two meetings between sessions of the plenary assembly. Table 4.1 presents the organization of the CCIR as established by its XIVth Plenary in 1978 and Table 4.2 presents the organization of the CCITT established by its Seventh Plenary in 1980.

---

[22] Ibid., Art. 11, 1. (3).

[23] As we will see, however, since 1956 the CCITT has always met at four-year intervals and since 1966 the CCIR has done the same.

[24] Ibid., Art. 58, 1.(a).

[24] Ibid., Art. 11, 4.

Each committee has a director, elected by the appropriate plenary assembly initially for a six year term (twice the official interval between two consecutive plenary assemblies) and thereafter for three year terms. At the time of writing, the Director of the CCIR is Richard C. Kirby of the United States and the Director of the CCITT is Leon Burtz of France. Each committee also has a specialized secretariat — in 1980 there were thirty-two individuals in the CCIR secretariat and forty-two in the CCITT secretariat.

As mentioned earlier, the CCITT also has its own laboratory staffed by the CCITT's specialized secretariat. It was created in 1927 to provide European administrations with a means for examining the transmission quality of subscriber telephone sets to determine if they met the international standards set by the CCIF for long distance telephone transmission.[26] The CCITT laboratory, now named the Telephonometric Laboratory, still tests telephone equipment for a fee. In 1980, for example, it was reported that eighteen subjective and objective tests were carried out for fifteen different administrations and private operating agencies from Western and Eastern Europe, Japan, and El Salvador for a total cost of 79,085.00 Swiss francs.[27] The laboratory also carries out research projects on "subjective and objective telephonometric problems" in conjunction with the CCITT's Study Group XII.

Participation in the work of the consultative committees as members is open to the administrations of all the members of the ITU and to private operating agencies which have been approved for participation by a member country. Participation in an advisory capacity is also open to "international organizations and regional telecommunication organizations . . . which coordinate their work with the International Telecommunication Union and which have related activities. . .". And "scientific or industrial organizations, which are engaged in the study of telecommunication problems or in the

---

[26] The original reference system was known as SFERT from the French translation of "European Telephone Transmission Master Reference System." Originally installed in Paris with equipment donated by the American Telegraph and Telephone Company, the laboratory was transferred to the ITU's Geneva headquarters as a result of the decisions of the 1947 Atlantic City Conferences. See Chapuis, *op. cit.*, p. 189.

[27] See ITU, Administrative Council, 36th Session, Geneva, June 1981, Doc. No. 5621-E (CA36-40), 31 March 1981, p. 5.4/33. This same document reports that the laboratory also purchased equipment worth 89,648.85 Swiss francs in the same period. *Ibid.*, p. 5.4/34.

design or manufacture of equipment intended for telecommunication services. . ."[28] and which have the approval of a member administration may be permitted to participate in study groups only in an advisory capacity. Representatives of private operating agencies can become chairmen and vice chairmen of working groups and, with authorization, represent member states. At the 1976 Plenary Assembly of the CCITT, for instance, of the eighteen study groups, the United States provided the chairmen for three, all of whom were employees of American Telegraph and Telephone company.[29]

As is usual with ITU conferences and meetings, the expenses of delegates are met by the administration or recognized private operating agency sending them. Other expenses of the meetings of plenary assemblies and working groups are met from the general ITU budget with the exception that the private operating agencies, international organizations, and regional telecommunications organizations all pay a pro-rated share of the expenses of all meetings to which they send representatives.[30]

The general procedure for considering questions is as follows. The plenary assembly draws up a list of questions to be investigated and assigns them to the study groups. Those administrations or private operating agencies which wish to take part in a particular group indicate their desire to the plenary assembly or the director of the CCI concerned. At the conclusion of the plenary assembly, the participants in the study groups take the questions home with them for examination. Questions are dealt with via correspondence under the directives of the study group chairman in the CCIR usually through a special rapporteur in the CCITT, and in meetings of members. Study group recommendations which reflect a substantial consensus of the participants are submitted to the next plenary

---

[28] In addition, the telecommunication operating service of the United Nations is entitled to participate in a consultative capacity. See *International Telecommunication Convention, Malaga-Torremolinos, 1973*, Art. 68. For a list of the private operating agencies, international and regional telecommunication organizations, and the scientific or industrial organizations which are permitted to participate in the consultative committees see Appendix No. 4.3.

[29] CCITT, Sixth Plenary Assembly, Geneva, 27 September - 8 October, 1976, *Orange Book*, Volume I, Geneva, 1977, pp. 217-221. The one other chair held by the United States — Special Autonomous Working Party (GAS 3) — was directed by an employee of General Telephone and Electronics (GTE).

[30] See below, Chapter 10.

assembly for final approval. The final texts, which are not signed by the delegates to the assembly or ratified by their administrations, are published by the secretariat of the ITU in the three working languages of the Union. In theory, they are simply statements of the committee's position on the subject. However, because they represent a consensus on the part of the key administrations, operating authorities, and major manufacturers, they enjoy considerable force and effect — often more so than regulations adopted by treaty.

Some questions have been the subject of continuing study over many years and a number of adjustments have been made to the pertinent recommendations.[31]

A recent innovation in procedure is the emergency device known as the "provisional recommendation," which was introduced because it was felt that technical advances sometimes demand solutions that cannot wait for the next plenary assembly. If adopted by unanimous agreement of the administrations and private operating agencies represented at a meeting of a study group, such recommendations can enter into force without awaiting the next plenary assembly's approval. When the plenary does meet, these provisional recommendations come before it for final approval in the normal way.[32]

In general the internal working methods of the two committees have developed along somewhat different lines. In the CCIR the study groups do most of their work in meetings of the full study groups. These run from one to three weeks and are well attended. The CCITT technique is to do much of its detailed work in many short meetings attended by a limited group of specialists for presentation in more or less finished form to their full study group meetings.

---

[31] M. Mili, "International Jurisdiction in Telecommunication Affairs," #5, *Telecommunication Journal,* Vol. 40, No. IX (September, 1973), pp. 562-566.

[32] M. Mili, "International Jurisdiction in Telecommunication Affairs," #6, *Telecommunication Journal,* Vol. 40, No. XII (December 1973), p. 746.

The CCITT at Work

Since its creation in 1956, the CCITT has held the following plenary assemblies:

- First, Geneva, 1956
- Second, New Delhi, 1960
- Third, Geneva, 1964
- Fourth, Mar del Plata, 1968
- Fifth, Geneva, 1972
- Sixth, Geneva, 1976
- Seventh, Geneva, 1980

The plenary assembly may be thought of as the starting place and at the same time the end point in the work of the committee as a unit. As the starting point, the plenary assembly organizes the committee for the following four-year interval, including the creation of the required study groups and working parties, selection of their chairmen and vice-chairmen, allocation of questions for study to the appropriate study group.

The plenary assembly is often faced with the problems which overlap the jurisdictions of two or more of its study groups, of the other consultative committee, or even of other international organizations. The only solution to such situations is to name a special rapporteur to act as a means of communication among all the affected groups or to rely on individuals who participate in each of the bodies which are concerned.

The plenary assembly then appoints the chairmen of the study groups and working parties. In order to qualify, they must be senior officials in their own administrations, they must have the ability to conduct meetings, and they must have a good knowledge of English or French.[33]

Finally, the plenary assembly has the task of electing the Director of the CCITT. Various factors must be taken into consideration when selecting a director. In particular, the director should not belong to

---

[33]P.J. Dormer, "The CCITT and its Secretariat," *Telecommunication Journal,* Vol. 32, No. 8 (August 1965) p. 316.

the same country as the Secretary-General or his deputy, or the director of the CCIR. In addition, consideration must be given to ensuring a wide geographical distribution of the various regions of the world among all the elected officials of the ITU.[34]

As the end point in the work of the committee, the plenary assembly has the task of approving the results of the work done during the preceding four-year period. Essentially, it approves a large number of texts, usually by unanimous decision. The CCITT, however, has the ability to reach and conclude final recommendations by correspondence. Again the study group involved must agree unanimously and twenty administrations must approve the result. When all of the texts have been edited by the secretariat of the CCITT, they are published in the three working languages of the Union. The entire set of texts is enormous (currently more than 6,000 pages) and contains all recommendations in force on a given subject as well as studies in progress on that subject.[35]

There is, however, another type of CCITT product known as "handbooks." Basically, these are condensed international experience put into subject-oriented form. They cannot be classified as design aids in the strict sense of the word, but provide instead a general knowledge of a given field of telecommunication.[36]

Contributions from administrations, recognized private operating agencies, and scientific or industrial organizations which are registered with a study group, working parties, and chairmen or rapporteurs must be sent to the Director of the CCITT at least three months before the date fixed for the opening of the meeting concerned. The director groups the documents according to the question involved, has the Secretariat make the necessary translations and sends them to the members of the study groups concerned.[37] These documents

[34] H. Baczko, "Results of the 23rd Session of the ITU Administrative Council," *Telecommunication Journal*, Vol. 35, No. 7 (July, 1968), p. 293.

[35] M. Mili, "International Jurisdiction of Telecommunication Matters." (December, 1973), pp. 748-749. For a further discussion on this subject, see below, Chapter Ten.

[36] G. Gosztony, K. Rahko and R. Chapuis, "The Grade of Service in the World-wide Telephone Network," *Telecommunication Journal*, Vol. 46, No. 10 (October, 1979), p. 627.

[37] CCITT, Seventh Plenary Assembly, Geneva, 10-12 November 1980, *Yellow Book*, Vol. I, Geneva, 1981, pp. 214-229.

are called "white documents". If, however, the contributions are not received in the given time limit, they are classified as delayed documents. Documents actually produced during the course of a meeting are designated "temporary documents."

Study groups generally try to avoid disagreement in discussing recommendations. If an objection is raised, the document is often continued for study or the problem is given to a special rapporteur to find a solution. If the results of the working parties meet the approval of the study group, a draft recommendation on the matter is sent to the secretariat of the CCITT which is responsible for its circulation to all the members registered in the study group.

Following the decision to eliminate most of the details of the Telegraph and the Telephone Regulations in favor of CCITT recommendations (as will be discussed in Chapter Nine), there has been a tendency to give those recommendations which must be put into effect at a certain date the appellation "instructions" to differentiate them from regular recommendations.[38]

The CCIR at Work

The plenary assembly of the CCIR has responsibilities that are similar to those of the plenary assembly of the CCITT. It decides on questions to be studied, forms working groups as needed and gives them their terms of reference, and elects its own director. It also designates the chairmen and vice chairmen of the study groups but, in contrast to the CCITT, it does not provide for more than one vice chairman per group. It establishes the schedule for working groups and other subordinate bodies and, when their work is done, it is in the plenary assembly where resolutions are passed and recommendations on standards are finally made.

CCIR plenary assemblies also provide the forum for establishing the technical data to serve as the basis for the work of Administrative Radio Conferences such as WARC-ST (1971), WARC-BS (1977) and WARC 79 (1979). Further, many of its studies facilitate the work of the IFRB.

---

[38] For a discussion on the problem see Y.M. Mao and E. Hammel, "Recommendation, Standard or Instruction?", *Telecommunication Journal,* Vol 48, No. XII (December, 1981), pp. 747-748.

There have been fifteen CCIR Plenary Assemblies:

1929	First	The Hague
1931	Second	Copenhagen
1934	Third	Lisbon
1937	Fourth	Bucharest
1948	Fifth	Stockholm
1951	Sixth	Geneva
1953	Seventh	London
1956	Eighth	Warsaw
1959	Ninth	Los Angeles
1963	Tenth	Geneva
1966	Eleventh	Oslo
1970	Twelfth	New Delhi
1974	Thirteenth	Geneva
1978	Fourteenth	Kyoto
1982	Fifteenth	Geneva

The work of the CCIR proceeds in a manner similar to that of the CCITT. The differences that exist are more structural and procedural than substantive. For example, the CCIR study group chairmen serve many of the same functions as do the rapporteurs in the CCITT, and the working party infrastructure is not as explict or as elaborate as that of the CCITT. The differences are largely historic in origin, but also tend to reflect the broader jurisdiction of the CCITT. The documents drawn up the the CCIR working groups are examined at regularly scheduled interim meetings and decisions are made whether to forward them to the plenary as reports or recommendations. At any point along this chain of decision-making, an administration may object to the results reached. If an objection is raised, the study is re-submitted to the working group and a representative of the objecting administration is asked to work with that group in order to clarify the purpose of the objection and to facilitate a solution to the problem. If agreement cannot be reached, the issue receives no recommendation and essentially fades out of the picture.

Participation

One of the aspects of the work of the consultative committees that has intrigued its observers over the years is the participation of individuals from private entities along with individuals from government agencies in important aspects of decision-making. While the logic of including manufacturers and private users of telecommunication equipment in the search for acceptable standards is strong, government entities are rarely willing to share international decision-making responsibilities so openly.

As of January 1, 1980, there were forty-eight recognized private operating agencies approved for participation in the work of the CCIR and forty-eight in the CCITT. Thirty-eight scientific or industrial organizations had approval to participate in the CCIR Study Groups and 135 in CCITT Study Groups. Thirty international organizations were also registered to work with the CCIR in an advisory capacity, and thirty to work with the CCITT.[39]

There are two major arenas which allow actual participation in the work of the consultative committees: the plenary assemblies and the study groups. Since the plenary assembly establishes the committee's work program and is ultimately responsible for the adoption of recommendations, it could be expected that it would attract more participants than any other single activity. Using the 1978 XIVth CCIR Plenary Assembly in Kyoto, Japan, as an example, twenty-one private operating agencies from nine countries were represented (except for two delegates, all others were included in the delegation of which they were nationals). Fourteen agencies in the "international organization" category sent observers, as did a UN Specialized Agency, the International Civil Aviation Organization (ICAO).[40] Table 4.2 lists the various participating entities.

At the CCITT's Sixth Plenary Assembly there were twenty-seven private operating agencies in attendance, twenty of which had their own separate listing and seven of which were reported as incorporated in the parent delegation. And thirteen international organizations sent observers, including two UN Specialized Agencies, the International Civil Aviation Organization, and the World Meterological Organization. Table 4.2 lists the entities involved.

There does not appear to be a marked increase in the participation of private operating agencies over the years, probably because most of the nations which have joined the ITU in the past three decades do

---

[39] See Annex E.

[40] See, CCIR, *Recommendations and Reports of the CCIR, 1978,* XIVth Plenary Assembly, Kyoto, 1978, Volume XIII-1, Minutes of the Plenary Sessions, Administrative Texts, Structure of the CCIR, Lists of CCIR Texts, Geneva, 1978, pp. 8-30.

not allow private enterprise in the telecommunication sector. The CCIR reports sixteen in attendance at its meeting in 1953 and the CCITT had seventeen at its plenary assembly in 1960. There has been an increase in the number of international organizations observing CCI work, although this increase has been uneven. In its 1953 Plenary Assembly the CCIR had six international organizations in attendance and at its 1960 plenary the CCITT had eight.

The ratio of non-government participants to government participants in the working groups is much higher in both the CCIR and the CCITT, with the honors going to the CCITT. The following is a breakdown of the CCIR study groups held during 1977:[41]

Study Group	Admins.	RPOAs	SIOs	IOs	% Non-admin.
1	30	13	5	1	39%
2	15	5	0	3	35%
3	22	8	3	0	24%
4	29	24	10	4	57%
5	29	16	0	3	40%
6	28	10	2	1	32%
7	20	4	0	2	23%
8	31	13	9	6	47%
9	32	17	15	1	51%
10	31	15	8	3	46%
11	32	19	7	3	48%
CMTT	32	19	10	2	49%
CMV	19	6	1	0	37%

The 1979 ITU Annual Report, on the other hand, provides the following profile of the participation of various administrations and non-administration entities in the work of the CCITT:[42]

---

[41] See ITU, *Report on the Activities of the International Telecommunication Union in 1977*, Geneva, 1978, p. 34.

[42] See ITU, *Report on the Activities of the International Telecommunication Union in 1979*, Geneva, 1980, pp. 57-62.

Study Group	Admin.	RPOAs	SIOs	IOs	% Non-admin.
I	23	15	11	2	55%
II	21	17	6	2	54%
III	21	22	3	6	60%
IV	11	13	12	1	70%
V	6	2	3	2	54%
VI	—	—	—	—	—
VII	17	15	27	7	74%
VIII	16	14	19	2	69%
IX	15	12	15	0	64%
X	15	13	1	0	48%
XI	22	14	30	0	67%
XII	17	10	7	0	50%
XIII	—	—	—	—	—
XIV	13	11	19	1	70%
XV	19	15	32	1	72%
XVI	16	11	12	0	59%
XVII	12	16	27	2	79%
XVIII	17	11	26	0	65%
CMBD	9	11	15	0	74%

It seems clear that the participation of private operating agencies and the manufacturers of telecommunications equipment in the establishment of international standards is more than merely symbolic. While national administrations tend to keep a closer rein on the non-governmental groups at the time of the final decision making in plenary assemblies, these groups are given a wide opportunity to affect the work of the study groups where the basic research takes place. It is interesting to note that the non-administration participants are a majority in only two of the CCIR study groups: that dealing with fixed service using communication satellites and the fixed service using radio-relay systems group. In 1979, in contrast, non-administration entities were in a majority in all of the CCITT study groups except for two, the study group on telegraph switching which was de-activated by the 1980 Plenary Assembly and the one on telephone transmission performance and local telephone networks.

The second most important characteristic of participation in the work of the consultative committees is the predominance of the developed nations, which is visible in the attendance at plenary assemblies but more striking in the working groups. At the 1979 CCIR plenary Assembly, nineteen out of twenty-seven of the OECD-related countries were represented, as were six out of the ten Warsaw Pact nations. Only thirty-six countries from the Third World were represented out of a possible 117 and only a few countries from the poorest Third World nations. The following fifteen countries sent delegations of over five in number which accounted for 200 of the total of 300 who attended: Canada (14); China (8); France (10); Germany (F.R.) (10); Indonesia (5); Italy (11); Japan (65); Korea (R.) (6); Nigeria (6); Norway (6); Spain (11); Switzerland (5); U.K. (15); U.S.A. (16); and the U.S.S.R. (12). Even without counting the large delegation of the host government, the predominance of developed nations is apparent.

The Third World made a much better showing at the 1976 Geneva CCITT Plenary, but all but one of the forty-nine missing delegations were from the lesser developed countries of the Third World. The following twenty-three national administrations sent delegations of five or more which amounted to 229 out of the total of 385 delegates in attendance: Japan (24); France (22); Italy (16); Canada (15); U.K. (14); U.S.A. (14); U.S.S.R. (14); Switzerland (13); Germany (F.R.) (9); Spain (8); Poland (8); Sweden (8); Belgium (7); China (7); Korea (R.) (7); Germany (D.R.) (6); Hungary (6); Netherlands (6); Australia (5); Denmark (5); Iran (5); Nigeria (5); and Norway (5).

In the study groups, the predominance of developed nations is most apparent. In 1977, for instance, of the thirty-nine countries which participated in the CCIR's thirteen study groups, only fourteen were from the Third World: Argentina, Bahrain, Brazil, China, Korea (R.), Cuba, India, Indonesia, Iran, Nigeria, Senegal, Uruguay, Venezuela, and Yugoslavia. Of the fourteen, only six attended more than half of the thirteen groups: China (13); India (13); Korea (R.) (13); Yugoslavia (12); Iran (8); and Indonesia (7). The countries with perfect attendance were primarily from the developed countries of the world: Australia, Canada, China, France, Germany (F.R.), India, Italy, Korea (R.), New Zealand, Nigeria, U.K., and U.S.A.[43]

In 1979, forty-four countries took part in the seventeen regular

---

[43] The U.S.S.R. participated in all but one study group and Finland, Poland, Spain, Sweden, and Switzerland participated in all but two.

CCITT study groups; of these, only fifteen could be classified as Third World, and only one had attended a majority of the study groups: Argentina (3); Brazil (6); China (6); Cyprus (1); India (1); Indonesia (2); Iraq (1); Jordan (2); Korea (R.) (1); Nigeria (8); Philippines (1); Tanzania (1); Thailand (1); Togo (1); and Uganda (1). Only six countries attended all seventeen study groups; all were devel- Fourteen additional countries attended a majority of groups: Austra- Sixteen additional countries attended a majority of groups: Australia (15); Austria (10); Belgium (13); Denmark (11); Finland (11); Germany (F.R.) (16); Hungary (10); Nigeria (8); Norway (16); Netherlands (16); Spain (15); Switzerland (16); U.S.A. (16); and U.S.S.R. (11).

Many reasons have been given for the failure of the developing countries to respond to the work of the CCIs. First, the developing countries lack the expertise necessary to participate effectively at a high technical level. Second, the developing countries lack the finances needed for participation. They cannot afford to send delegates to study group meetings, especially in view of the fact that there are so many to attend, nor can they spare the competent personnel at home to follow developments in the two committees. The situation has created a vicious circle: without participation by developing countries, the study groups and committees have found it difficult to identify satisfactorily questions and problems of specific concern to these countries.

A number of solutions have been put forth. More serious efforts could be made to encourage participation, such as holding meetings in developing countries. There have also been suggestions to create small teams of experts from the CCITT and the CCIR to be made available to the developing nations. Although they may seem plausible, there remain major drawbacks to these solutions. In the first place, as concerns the small team of experts, it is not in the purview of the CCIs to be "operationally" functional; this is within the boundaries of the technical cooperation section of the ITU secretariat. Although the CCITT and CCIR have organized seminars at times to explain the work of the study groups, more frequent seminars might help to increase the level of understanding among experts from developing countries and lead to the confidence needed for effective participation. Moving the CCITT/CCIR meetings to the developing countries remains beyond the budget of the ITU to transport the required number of translators and interpreters and records from one part of the world to another, and it would not help to resolve the principal problem of a lack of experienced personnel.

Some success has been achieved through holding seminars and publishing handbooks on special problems of the developing countries, but the problem is still a serious one which, if unresolved, may have repercussions on the work of the ITU.

Conclusions

The two surviving international consultative committees are unique in the ITU experience, and in the experience of most other international organizations, because of their participatory blend of government and private enterprise, the organizational independence they enjoy within the overall institutional structure, and their freedom in choosing problems for investigation. Over the years these three characteristics have helped them gain a high reputation among the developed nations of the world for the quality of their recommendations.

The reputation of the CCIs is not as strong among the developing countries, however. The immediate, perceived needs of the LDC's are not for the most advanced— and often the most expensive — technologies with which the CCIs have traditionally been involved, but for intermediate and less expensive technologies more attuned to systems in which the finances for telecommunications are much more difficult to obtain. For this reason, many developing countries' administrations tend to refrain from active participation in the real work of the CCIs. The problem is compounded by the fact that many of these administrations also lack the finances and the trained personnel to devote to the time-consuming activities of the CCIs.

In effect, a large part of the membership of the ITU is denied participation in two of its major organs. Under the circumstances, it seems clear that for the consultative committees to continue to provide a forum for setting standards for new and developing telecommunication technologies, something should be done either to convince the Third World majority that the work of the CCI's is important to them and the future of their telecommunication networks, to provide more time for the study of problems of special interest to the third world, or to arrange for the active participation of more Third World administrations. After all, since the CCI's are only sub-organs of a universal international organization, their future would seem to depend quite heavily on the goodwill of all, or at least a vast majority, of their clientele.

TABLE 4.1*
CCIR AND CCITT STRUCTURE
I. CCITT STUDY GROUPS AND OTHER BODIES (1981 - 1984)

A. Study Groups

I. Definition and operational aspects of telegraph and telematic services (facsimile, telex, videotex, etc.)
II. Telephone operation and quality of service
III. General tariff principles
IV. Transmission maintenance of international lines, circuits and chains of circuits; maintenance of automatic and semi-automatic networks
V. Protection against dangers and disturbances of electromagnetic origin
VI. Protection and specifications of cable sheaths and poles
VII. Data communication networks
VIII. Terminal equipment for telematic services (facsimile, telex, videotex, etc.)
IX. Telegraph networks and terminal equipment
X. (Not in use 1981-1984)
XI. Telephone switching and signaling
XII. Telephone transmission performance and local telephone networks
XIII. (Not in use 1981-1984)
XIV. (Not in use 1981-1984)
XV. Transmission systems
XVI. Telephone circuits
XVII. Data communication over the telephone network
XVIII. Digital networks

B. Joint Working Parties

GM LTG Use of telephone-type lines for purposes other than telephony. CCITT Working Groups involved: IV, VIII, IX, XV, XVIII, and CMBD. (Working Group XV controlling.)

GM SMM Maritime mobile service. CCITT Working Groups involved: I, II, and III. (Working Group II controlling.)

---

*From *Telecommunication Journal* Vol. 48, No. IV (April, 1981). pp. 184-188 and ITU, *Report of the CCIR, 1978*, pp. 170-175 and 237-246.

C. Special Autonomous Groups

GAS 4 Primary sources of energy

GAS 5 Economic and social problems relating to telecommunication development

GAS 7 Rural telecommunication

GAS 8 Economic and technical impact of implementing a regional satellite network

GAS 9 Economic and technical aspects of transition from an analog to a digital telecommunication network

D. Regional Tariff Groups of Study Group III

GR TAF   Africa
GR TAL   Latin America
GR TAS   Asia and Oceania
GR TEUREM Europe and the Mediterranean Basin

II. CCIR STUDY GROUPS AND OTHER BODIES (1978-1982)

A. Study Groups

1. Spectrum utilization and monitoring
2. Space research and radioastronomy
3. Fixed service at frequencies below about 30 MHz
4. Fixed service using communication satellites
5. Propagation in non-ionized media
6. Ionospheric propagation
7. Standard frequencies and time signals
8. Mobile services
9. Fixed service using radio-relay systems
10. Broadcasting service (sound)
11. Broadcasting service (television)

B. Interim Working Parties

Plen/1. Review of the work of the CCIR

Plen/3. Updating the texts of the Special Report on possible broadcasting satellite systems and their relative acceptability

1/2   Use of analysis techniques and computers in frequency management

4/1   Technical considerations affecting the efficient use of the geostationary orbit

5/1   Prediction of phase and amplitude of ground-waves

5/2   Tropospheric propagation data for broadcasting, space, and point-to-point communications

5/3   Influence of the non-ionized regions of the atmosphere on wave propagation

6/1   Sky-wave field strength and transmission loss at frequencies above 1.6 MHz

6/2   Radio noise

6/3   Basic long-term ioniospheric predictions

6/4   Sky-wave propagation at frequencies between 150 and 1600 kHz

6/5   Propagation at frequencies below about 150 kHz with particular emphasis on ionospheric effects

6/7   Operational parameters for ionospheric radio circuits

6/8   VHF propagation by sporadic E

6/9   Ionospheric factors influencing communication and navigation systems involving spacecraft

6/10 Ionospheric modification by high-power transmissions

6/11 Special problems of radiocommunication associated with the high-latitude ionosphere

7/4   World-wide time dissemination by means of satellites

7/5   Inaccuracy and reliability of frequency standards and reference clocks

8/5   Automated VHF/UHF mobile radiotelephone systems

8/6   Numerical identification of ship stations

8/7   Technical and operating characteristics of systems in the maritime mobile satellite service

11/2 Protection ratios for the broadcasting-satellite service for the purpose of frequency sharing (television)

11/3 Broadcasting services intended for alphanumerical and/or graphic display

CMTT/1 Digital systems for the transmission of sound-programme and television signals

CMV/1 Terms and definitions

III. JOINT CCIR/CCIT STUDY GROUPS AND OTHER BODIES

A. Administered by the CCIR

1. Joint Study Groups

CMTT Television and sound transmission

CMV Definitions and symbols

B. Administered by the CCITT

1. Joint Study Groups

CMBD Circuit noise and availability

2. Joint Plan Committees

World Plan — General plan for the development of the world telecommunication network

Africa Plan — General plan for the development of the regional telecommunication network for Africa

Latin American Plan — General plan for the development of the regional telecommunication network in Latin America

Asia Plan — General plan for the development of the regional telecommunication network in Asia and Oceania

Europe Plan — General plan for the development of the regional telecommunication network in Europe and the Mediterranean Basin

3. Special Autonomous Groups

GAS 3 — Economic and technical aspects of the choice of transmission systems

Decision-Making Structure

TABLE 4.2
PARTICIPANTS AT THE SIXTH PLENARY ASSEMBLY
OF THE CCITT IN GENEVA IN 1976
I. MEMBER ADMINISTRATIONS

1. Algeria	35. Greece	70. Niger
2. Argentina	36. Honduras	71. Nigeria
3. Australia	37. Hungary	72. Norway
4. Austria	38. Iceland	73. Oman
5. Bangladesh	39. India	74. Pakistan
6. Belgium	40. Indonesia	75. Panama
7. Benin	41. Iran	76. Philippines
8. Bolivia	42. Iraq	77. Poland
9. Brazil	43. Ireland	78. Portugal
10. Bulgaria	44. Israel	79. Qatar
11. Byelorussia	45. Italy	80. Roumania
12. Cameroon	46. Ivory Coast	81. Saudi Arabia
13. Canada	47. Japan	82. Senegal
14. Central African	48. Jordan	83. Somali
Republic	49. Kenya	84. Spain
15. Chad	50. Korea (D.R.)	85. Sweden
16. Chile	51. Korea (R.)	86. Switzerland
17. China	52. Kuwait	87. Syria
18. Colombia	53. Laos	88. Tanzania
19. Congo	54. Lebanon	89. Thailand
20. Costa Rica	55. Liberia	90. Togo
21. Cuba	56. Libya	91. Tunisia
22. Cyprus	57. Liechtenstein	92. Turkey
23. Czechoslovakia	58. Luxembourg	93. Uganda
24. Denmark	59. Madagascar	94. U.K.
25. Ecuador	60. Malaysia	95. Ukraine
26. Egypt	61. Mali	96. United Arab
27. El Salvador	62. Mauritania	Emirates
28. Ethiopia	63. Mauritius	97. U.S.A.
29. Finland	64. Mexico	98. U.S.S.R.
30. France	65. Monaco	99. Upper Volta
31. Gabon	66. Morocco	100. Venezuela
32. Germany (D.R.)	67. Mozambique	101. Yemen
33. Germany (F.R.)	68. Netherlands	102. Yugoslavia
34. Ghana	69. New Zealand	103. Zaire

II. Recognized Private Operating Agencies*

1. Austria
   Radio Austria A.G.

2. Brazil
   Empresa Brasileira de Telecommunicacoes S.A.

3. Canada
   Canadian Telecommunications Carriers Association (CTCA)

4. Denmark
   The Great Northern Telegraph Co.

5. Nigeria
   Nigeria External Telecommunications Ltd.

6. Philippines
   Philippine Global Communications, Inc.
   Philippine Long-Distance Telephone Company, Inc.

7. Switzerland
   Radio-Suisse S.A.

8. U.K.
   Cable and Wireless Ltd.
   International Marine Radio Company Ltd.
   The Marconi International Marine Company Ltd.

9. U.S.A.
   American Telephone and Telegraph Co. (A.T.&T)
   Communications Satellite Corporation (COMSAT)
   General Telephone and Electronics Corporation (GTE)
   ITT World Communications, Inc.
   RCA Global Communications, Inc.
   Telenet Communications Corporation
   TRT Telecommunications Corporation
   Western Union International, Inc.
   The Western Union Telegraph Company

---

*Seven additional private operating agencies were reported as integrated into national delegations: Canada (1) Italy (3), Japan (2), and Spain (1).

III. International Organizations

1. European Computer Manufacturers Association (ECMA)
2. International Air Transport Association (IATA)
3, International Chamber of Commerce (CCI)
4. International Chamber of Shipping (ICS)
5. International Marine Radio Association (CIRM)
6. International Electrotechnical Commission (CEI)
7. International Conference on Large High Tension Electrical Systems (CIGRE)
8. International Press Telecommunications Council (IPTC)
9. European Organization for Nuclear Research (CERN)
10. International Organization for Standardization (ISO)
11. Arab Telecommunications Union

IV. UN Specialized Agencies

1. International Civil Aviation Organization (ICAO)
2. World Meteorological Organization (WMO)

PARTICIPANTS AT THE XIVth CCIR PLENARY ASSEMBLY, KYOTO, JAPAN, 1978

I. Countries

1. Argentina	21. Indonesia	41. Oman
2. Australia	22. Iran	42. Pakistan
3. Austria	23. Iraq	43. Panama
4. Bangladesh	24. Ireland	44. Papua New Guinea
5. Brazil	25. Italy	45. Philippines
6. Canada	26. Ivory Coast	46. Poland
7. China	27. Japan	47. Roumania
8. Cyprus	28. Kenya	48. Spain
9. Czechoslovakia	29. Korea (R.)	49. Sri Lanka
10. Cuba	30. Kuwait	50. Saudi Arabia
11. Denmark	31. Madagascar	51. Sweden
12. Egypt	32. Malaysia	52. Switzerland
13. Finland	33. Mali	53. Tanzania
14. France	34. Mauritius	54. Thailand
15. Germany (D.R.)	35. Mexico	55. Togo
16. Germany (F.R.)	36. Morocco	56. Turkey
17. Ghana	37. Netherlands	57. U.K.
18. Guinea	38. New Zealand	58. U.S.A.
19. Hungary	39. Nigeria	59. U.S.S.R.
20. India	40. Norway	60. Venezuela
		61. Zaire

II. Recognized Private Operating Agencies

1. Canada
   a. Canadian Broadcasting Corporation
   b. Canadian Telecommunications Carriers Association
   c. Teleglobe Canada (Ex-COTC)

2. Germany
   a. Deutsche Welle
   b. Norddeutscher Rundfunk
   c. Zweites Deutsches Fernsehen

3. Italy
   a. Intalcable
   b. Rai-Radiotelevisione Italiana
   c. Telespazio S.P.A.

4. Japan
   a. Kokusai Denshin Denwa Co. Ltd. (KDD)
   b. Nippon Hoso Kyokai (NHK) (Broadcasting Corporation of Japan)
   c. Nippon Minkan Hoso Renmei (National Association of Commercial Broadcasters of Japan)
   d. Nippon Telegraph and Telephone Public Corporation (NTT)

5. Kenya
   a. Kenya External Telecommunications Co. Ltd.

6. Nigeria
   a. Nigerian External Telecommunications Ltd.

7. Spain
   a. Compania Telefonica Nacional de Espana

8. U.K.
   a. British Broadcasting Corporation (BBC)
   b. Cable and Wireless Ltd.
   c. Independent Broadcasting Authority (IBA)
   d. United Kingdom Post Office

9. U.S.A.
   a. American Telephone and Telegraph Co. (AT&T)
   b. Communications Satellite Corporation (COMSAT)

III. International Organizations

1. African Postal and Telecommunications Union (UAPT)
2. Committee on Space Research (COSPAR)
3. European Broadcasting Union (EBU)
4. European Space Agency (ESA)
5. International Air Transport Association (IATA)
6. International Association of Lighthouse Authorities (IALA)
7. International Astronautical Federation (IAF)
8. International Astronomical Union (IAU)
9. International Maritime Radio Association (CIRM)
10. International Radio and Television Organization (OIRT)
11. International Union of Radio Science (URSI)
12. International Telecommunications Satellite Organization (INTELSAT)
13. Inter-Union Commission Allocation of Frequencies for Radio Astronomy and Space Science (IUCAF)

IV. U.N. Specialized Agency

International Civil Aviation Organization (ICAO)

chapter 5

THE INTERNATIONAL FREQUENCY REGISTRATION BOARD

Of the two major changes in the structure of the ITU made by the 1947 Atlantic City Telecommunication Conference, the International Frequency Registration Board was potentially the most innovative. Some of its originators actually foresaw a body of eminent technicians resolving disputes among administrations over the use of radio frequencies. The subsequent failure of the member administrations to provide the new engineered frequency list which would have been the base for its activities, substantially undercut the role of the IFRB to the point where it almost had no reason to continue to exist. Since that time, however, the IFRB has found a way to reinforce and expand its mandate to the point where it has a significant role in the affairs of the ITU and its member states that makes it almost invulnerable to serious attack.

Background

The IFRB did not come into existence until 1947, but the problem that it was created to solve dates back to the early days of radio communication: the possibility of interference between radio stations operating on or near the same frequency. Governments initially attempted to overcome this problem through bilateral arrangements. Such efforts were only partially successful. As radio increased its range and more countries began using it, international action became a necessity. Based to a certain extent on the Final Protocol produced three years earlier, the Berlin Radiotelegraph Conference of 1906 adopted a number of provisions intended to reduce the possibility of interference between maritime stations, including methods of organizing radio traffic, limiting transmitter power, and eliminating unnecessary messages. It also allocated a number of channels for use in the maritime service and each administration was requested to forward to the International Bureau of the International Telegraph Union information for publication concerning the ship and coastal stations operating under its authority, including call letters, hours of operation, radio systems used, location of coastal stations, and the frequencies used. This information was available to administrations inaugurating a new radio service or modifying an old one. These arrangements were left largely untouched by the 1912 International Radiotelegraph Conference.

The rapid development of radio communication technology during World War I, along with a subsequent increase in the number of radio stations engaging in an expanding variety of international communications services, such as broadcasting, aircraft radio, and long range communication necessitated the calling of a new conference. At this conference, the 1927 Washington International Radiotelegraph Conference, the concept of providing information on frequency usage was carried one step further. Although the delegates refused to endorse an explicit legal notion of priority for prior use, they did agree to a more implicit one —that all frequencies for radio stations should "be chosen so as to avoid so far as possible interference with international services carried on by existing stations the frequencies of which have already been notified to the International Bureau."[1]

With continued problems of interference, only partially solved on a regional basis, the Madrid Radio Conference of 1932 flirted with the notion of a legally explicit concept of priority for prior use, but once again it was decisively defeated. The conference did, however, reinforce the notification procedure by requesting that administrations notify the Bureau before new frequencies were placed into use or the use of old ones was modified.

The period 1932 to 1947 was a time of rapid advances in technology and a further rapid increase in the use of radio. By 1947 a great number of stations were operating on frequencies that had never been registered with the Berne Bureau and others were operating on frequencies previously employed by other countries before the war had disrupted their services. This chaotic situation led the members of the ITU to investigate two interrelated changes: a new engineered

---

[1] As quoted in Codding, *The International Telecommunication Union,* p. 126.

frequency list to replace the old Berne list and an International Frequency Registration Board to approve modifications to the new list and to help the ITU in the area of international frequency management.

The originators of the IFRB concept thus proposed a marked innovation in the functioning of international organizations. As stated by Harold K. Jacobson, the Americans saw the new organ of the ITU "as something of a cross between the Federal Communications Commission and the International Court of Justice."[2]

The idea of a strong IFRB soon ran head-on into the older and stronger concept of national sovereignty, however. When translated into telecommunication terms, the latter concept meant that national administrations had the power to decide for themselves what frequencies they wished to use. While the IFRB was to examine all new notifications to the new engineered international frequency list and changes in frequency use for compatibility with the relevant rules and regulations and the possibility of creating harmful interference, the only discretionary action available to it would be to indicate in the list whether such notifications were or were not so compatible. If an administration decided to proceed with the use of a frequency that the IFRB judged contrary to the rules for the use of frequencies, there was nothing that it could do. Further, if a dispute should occur over harmful interference, if requested by an administration, the IFRB could help the disputants come to a mutual arrangement, but it was given no powers whatsoever to make a determination on the legal merits of the case. In other words, the IFRB was to help and suggest, but not to decide.

A good description of the functions that were given to the IFRB by the Atlantic City Radio Conference is contained in the statement made by the chairman of the working group that drew up the board's procedures. The role to be played by the Board, he stressed, "was that of *witness* and *nothing more*":

"In the matter of the registration of frequency allocations, the I.F.R.B. will base its decisions solely upon technical considerations. The nature of these considerations is determined very

---

[2] Harold K. Jacobson, "The International Telecommunication Union: ITU's Structure and Functions," p. 49.

clearly ... The I.F.R.B will function here as a verification Board, such as Lloyd's of London. It will judge according to codified standards, equally applicable to all frequency allocations, no matter what their source. If the allocation of a frequency is perfectly correct with respect to I.F.R.B Regulations, such allocation, if I may so express myself, will be stamped 'registration.' If it is not perfectly correct, it will be stamped 'notification.' *And that is all.*

The stamp 'notification' would, moreover, in no way prevent the country making the assignment so classified from putting it to use. Only later, should clashes arise, would the importance of the stamp be recognized insofar as the arbitrators take it into consideration. It is obvious, in this respect, that the value of the classification will depend largely upon the reputation for skill and excellence in making decisions earned by the I.F.R.B."[3]

The IFRB was given two other major tasks by the Atlantic City conferences. It was given the power to study problems dealing with international frequency management when so requested by administrations and it was given the power to cancel frequency assignments that were not placed into operation within two years following the date of receipt of the first notice, or that had been out of use for a period of three years, but only if it obtained the agreement of the administrations which had notified them.[4]

It is obvious that the IFRB, as created in 1947, was far from being an international Federal Communications Commission or an International Court of Justice in frequency matters. In any event, eleven countries were chosen by the Radio Conference to send representatives to the Board and soon it was actively involved in helping to prepare the new frequency list.

The first crisis in the life of IFRB came when the conferences that had been scheduled by the Atlantic City Radio Conference to draft

---

[3] As quoted in Codding, *The International Telecommunication Union,* p. 245.

[4] As regards these tasks, the chairman of the working group that drew up the Board's procedures pointed out that "no initiative of its own is left to the I.F.R.B. This is, he continues, "a further consideration of a reassuring nature for countries which persistently display anxiety over the possibility of some dictatorial action on the part of the International Frequency Registration Board." Ibid., p. 246.

the new international frequency list, engineered to the needs of the nations of the world, ended in chaos. With only a few minor exceptions, it was proven impossible to come to any agreement after more than three years of intensive effort. In addition to some political problems, the efforts failed primarily because the frequency requirements submitted by member countries far exceeded the available bandwidth.

In order not to jeopardize the development of the various radio services, the 1951 conference gave the IFRB an important role to play in bringing into force various provisions of the 1947 Radio Regulations, including the move of radio stations to their new service allocations.

By 1959 it had become clear that a new engineered frequency list was unattainable. The 1959 Administrative Radio Conference therefore issued a series of directives which modified the IFRB's registration procedures, extended them to include the greater part of the frequency spectrum where no plans had been concluded, and decided to create a new master frequency register to which the older recorded assignment could be transferred.

In 1965 the IFRB faced its second major crisis. A number of proposals were submitted to the 1965 Montreux Plenipotentiary Conference to abolish the Board and give its ongoing duties to the General Secretariat. The reasoning given was that since the IFRB had completed all of the important, but temporary, tasks that had been given to it by the Atlantic City Conferences and the Radio Conferences of 1951 and 1959, it had become simply an agency for the registration of frequencies. It would be much less costly to give this task to a division of the General Secretariat which could be just as impartial and independent. The IFRB, in effect, had outlived its usefulness.

The IFRB might well have been abolished in 1965 had it not been for a new development. The influx of new,developing nations to ITU membership had begun by this time and their representatives looked on the IFRB as an important source of advice on frequency management problems and, rightly or wrongly, as a potential ally in dealings with the more developed nations in conflicts over frequency usage. In effect, this idea appealed to enough delegations to save the IFRB from extinction. The high cost of an eleven man board was a telling one, however, so it was relatively easy for the 1965 plenipotentiary to agree to cut the number to five.

There have been no attempts to eliminate the IFRB since 1965. The activities of the Board in the registration of radio frequencies, supplemented more recently by tasks of preparing draft allotment plans for subsequent adoption at administrative conferences, holding frequency management seminars, and the perception of the Board as a protector of the interests of the developing nations, has given it a strong constituency. This perception arises from the many provisions that have been added to the Radio Regulations over the past twenty years which impose an affirmative obligation on the Board to give special assistance to developing countries. The view was greatly advanced by a decision of the WARC 79 Conference which actually requires preferential treatment by the Board of request from developing nations for certain high frequency channels.

Functions

The essential duties of the IFRB are set forth in Article 10 of the International Telecommunication Convention and Articles 9 to 15 of the Radio Regulations. As summarized by the IFRB itself, they are:

(1) To effect an orderly recording of frequency assignments made by the different countries and of the positions assigned by them to their geostationary satellites in accordance with the procedures prescribed in the Radio Regulations; and

(2) To furnish advice to members of the Union with a view to the operation of the maximum practical number of radio channels in those portions of the spectrum where harmful interference may occur and to the equitable, effective, and economical use of the geostationary satellite orbit.[5]

Elaborating on the first of these two tasks, again in the words of the IFRB:

---

[5] See ITU, IFRB, *Frequency Management and the Use of the Radio Frequency Spectrum and of the Geostationary Satellite Orbit,* Geneva, 1979 (mimeo.), p. 5.

". . . it can be said that one of the IFRB's main tasks is to decide whether radio frequencies which countries assign to their radio stations are in accordance with the Convention and the Radio Regulations and whether the proposed use of the frequencies concerned may cause harmful interference to other radio stations which are already in operation. Thus, the Board determines—on purely technical bases—the right of an Administration to use a given frequency for a specific purpose and the responsibilities the Administration thereby assumes vis-a-vis other Administrations."[6]

What the Board refers to as its primary task is the procedure for recording notification of frequencies from member administrations in the Master Frequency Register and the legal consequences, undefined as they are, which flow from such registration. The procedure is as follows:

1. Administrations notify the IFRB of any frequency assignment or change in the basic characteristics of an existing assignment if:

   a. The use of the frequency is capable of causing harmful interference to any service of another administration;

   b. The frequency is to be used for international communication; or

   c. It is desired to obtain international recognition of the use of the frequency.

2. The IFRB makes an examination of the frequency assignment to see if it could cause harmful interference to a station of another administration and whether the notification is in general conformity with the Table of Frequency Allocations and other provisions of the Radio Regulations.

3. If the IFRB judges that there is little or no probability of harmful interference and that it is in conformity with the Table of Frequency Allocations and other provisions of the Radio Regulations, it enters the assignment in the Master Frequency Register. If the assignment is not in order it will be returned by the IFRB and normally that administration will find an alternative frequency for the service in question. If the administration insists, the

---

[6] Ibid.

assignment will be recorded in the Register even without a favorable examination if it can be shown that it can be used without causing harmful interference. Registration of a frequency in the Register bestows on the notifying administration "a right to formal international recognition" and/or "the right to international protection from harmful interference" although nowhere are those terms defined.

4. The information contained in the Master Frequency Register is published periodically in the International Frequency List which is distributed to member administrations.[7]

The processing of frequency notifications from member administrations is a not inconsiderable task. In 1979, for instance, the IFRB reported that it had given "full technical examination" to 52,304 frequency assignments.[8]

As the IFRB has been given additional tasks, including preparations for more and more conferences, it has had increased difficulty in keeping up with the workload. The Administrative Council was told at its 1981 meeting that the IFRB had a backlog of some 84,400 assignments.[9] This situation has led the Administrative Council to a number of discussions of the IFRB's working methods and, recently, an authorization to increase the size of the IFRB specialized secretariat.

The IFRB has two related duties. Each year the IFRB publishes the four Tentative High Frequency Broadcasting Schedules together with recommendations for improving the utilization of that sector of the frequency spectrum. This procedure was adopted to aid administrations in planning their HF broadcasting services when it was found impossible to come up with a new allotment plan. The number of frequency-hours included in each of the HF broadcasting schedules was approximately 24,750 in November, 1979.

---

[7] By the end of 1979 the Master Register contained details concerning about 563,000 frequency assignments. See ITU, *Report on the Activities of the International Telecommunication Union in 1979*, Geneva, 1980, p. 22.

[8] Ibid.

[9] U.S., Department of State, U.S. Mission in Geneva to the Secretary of State, International Telecommunication Union: 36th Session of Administration Council (Geneva, 1-19 June 1981), Incoming Telegram No. 5769, June, 1981.

The second related task is to coordinate the advanced planning and the putting into operation of satellite communication systems. This is a relatively new and complicated procedure that includes among other things both the frequencies to be used and the orbital position of the satellite. The IFRB received advance information from administrations concerning twenty-five new satellite networks in 1979. This function is expected to increase as more satellites are placed in orbit.

There is one additional function performed by the IFRB which is related to the recording of frequency assignments, and that is the task of reviewing assignments for actual usage. If it appears to the Board that a recorded assignment has not been brought into regular operation as notified, it may cancel the notification with the agreement of the notifying administration. In 1979, for example, the IFRB's review of entries in the Master Frequency Register in the bands below 28 MHz resulted in 10,878 modifications and 7,485 cancellations. In bands above 28 MHz there were 1,061 modifications and 4,178 cancellations.

The IFRB performs a number of other tasks as determined by other ITU organs. Among the more important of these, at least in the past twenty years, has been aiding in the technical planning for administrative radio conferences. The IFRB, for instance, did a great deal of preparatory work for the 1979 World Administrative Radio Conference, including a study of proposals to that conference and the preparation of working documents, and the 1979 conference, in turn, gave the IFRB a number of tasks dealing with technical studies for the series of follow-up conferences that it set in motion. More recently, it was involved in the preparation for the Regional Administrative MF Broadcasting Conference for Region 2 held in Buenos Aires in 1980.

The IFRB has also become increasingly involved in activities which can be classified as development assistance. As mentioned earlier, the IFRB is helping more and more administrations with their frequency management problems, including advice on the choice of frequencies which can best provide for their needs. The IFRB is also holding periodic seminars on frequency management and the use of the radio frequency spectrum. These seminars had their origin in a realization that the provisions of the Radio Regulations that deal with the notification and registration of frequency assignments were becoming increasingly complex and hard to understand. The first such seminar was held in 1962 at the IFRB's initiative. Its value was recognized by the Administrative Council which gave its official blessing by Resolution No. 528 of April, 1963. Seminars under this authorization were offered in 1963, 1964, 1966, 1968, 1970, 1972, 1974, 1976, 1978, and 1980.

Basically, an IFRB seminar consists of a number of speeches by members of the IFRB, its secretariat, and others from the ITU or outside who volunteer to talk about the work of the ITU, the IFRB (the notification and registration of frequencies in particular), developments in the use of various frequencies, and the like. Usually two weeks in duration, some time is set aside on the two Fridays for questions and discussion. Simultaneous interpretation is provided in English, Spanish, French, Russian, and Chinese, and all documents are translated into the three official working languages.

Participation in the IFRB seminars has continued to grow. There were seventy-one participants from forty-four administrations at the first seminar and 209 participants from eighty-five administrations in 1980. UNDP funds have been used to increase the number of participants from the developing countries. In 1980, for instance, thirty-three participants received UNDP fellowships.[10]

Composition

The qualifications for serving on the IFRB tend to reflect to a great extent the orientation of the delegates at Atlantic City who anticipated the creation of a semi-judicial organ. The IFRB, according to the Malaga-Torremolinos Convention, "shall consist of five independent members" who "shall serve, not as representing their respective countries, or of a region, but as custodians of an international public trust."[11]

The members of the Board are protected from national pressures by provisions similar to those which apply to most other international civil servants in the UN system, whereby member governments are

---

[10] See ITU, Administrative Council, 36th Session, Geneva, June 1981, Addendum No. 6 to Document No. 5621-E, 28 May 1981, p. 5.2/8.

[11] ITU, *International Telecommunication Convention, Malaga-Torremolinos, 1973*, Art. 10, paras. 1 & 2.

forbidden from attempting to influence them in their duties. As an additional effort to insure the independence of Board members in the performance of their duties, governments with a member on the Board are requested to "refrain, as far as possible, from recalling that person" during his term of office.[12]

The IFRB has been remarkably free from accusations of national partisanship. The U.S.S.R., however, has tended to ignore the admonition against recalling members during their terms of office. The U.S.S.R. has provided eight members of the Board; the longest any one has served is seven years. The average term of the Russians has been 4.4 years. Two members of other nationalities served for 28 years and the average length of service of the non-U.S.S.R. nationals has been 12.3 years. See Table 5.1 for the names and nationalities of the members of the IFRB and their length of service.

On the matter of personal qualifications, the convention provides that members of the Board "shall be thoroughly qualified by technical training in the field of radio and shall possess practical experience in the assignment and utilization of frequencies." Moreover, "each member shall be familiar with geographic, economic, and demographic conditions within a particular area of the world."[13]

Over the years the members of the Board have clearly fulfilled the first of the two qualities required of them. As far as can be determined, all have been trained in the field of radio and have had experience in the assignment and utilization of frequencies in governmental agencies. Looking at the qualifications of the five members of IFRB at the time of writing, for example, Abderrazak Berrada (Morocco) is the recipient of a Diploma in Engineering from the Ecole Superieure d'Electricite in Paris, specializing in radioelectricity and electronics. He had several posts in government, the latest of which was Secretary-General of the Morocco Posts, Telegraph, and Telephone Ministry. Sakae Fujiki (Japan) obtained an engineering degree with specialization in electrical engineeering and, before becoming a member of the Board, served as Director-General of the Radio Regulatory Bureau in Japan's Ministry of Posts and Telecommunications. Petr Sergeevich Kurakov (U.S.S.R.)

---

[12] Ibid., Art. 57, para. 2. (5).
[13] Ibid., Art. 57, para. 1.

received his education at the Moscow Aviation Institute as a space radiocommunication engineer and served as Chief of Division of the Main Space Communications Department of the Russian Ministry of Telecommunications. Francis Gerard Perrin (Canada) is an electrical engineer who served as Head of the International Agreements and Frequency Notification Unit of the Canadian government. Finally, Charles William Sowton (U.K.) is an electrical engineer and served as the Director of the British Directorate of Radio Technology in the Ministry of Posts and Telecommunications.[14]

In 1980 the average age of the five members of the IFRB was 56.4 years. The youngest was Kurakov at 44 and the oldest was Sowton at 70.

The influential specialized secretariat of the IFRB is divided into a Regulatory Department and an Engineering Department. The largest is the Regulatory Department, which is composed of three divisions; Frequency Registration, Publications and Administrative Division; Regulations Application Division; and the Coordination and Agreements Division. The Engineering Department has five divisions: the Standards and Procedures Division; the Space Services Division; the Fixed Service Division; the Mobile Services Division; and the Broadcasting Service Division.[15] As the end of December 1979, there were 98 employees in the IFRB specialized secretariat, thirty-seven in the professional category and above, and the remainder in the general service category.[16]

Elections

The convention provides that the five members of the Board be elected by the plenipotentiary conferences from candidates "sponsored" by member countries, "in such a way as to ensure equitable distribution amongst the regions of the world".[17] No member may nominate more than one of it own nationals.

The elections to be held in 1982 will be the first held under this

---

[14] From curriculum vitae provided by members at the time of election and supplied to the authors by the IFRB Secretariat.

[15] For an account of the organization of the IFRB specialized secretariat see ITU, Administrative Council, 34th Session, June 1979, Doc. No. 64-E, p.4.

[16] See ITU, Administrative Council, 35th Session, May 1980, Addendum No. 2 to Document No. 5468-E (5 March 1980).

[17] ITU, *International Telecommunication Convention, Malaga-Torremolinos, 1973*, Art. 10, para. 1.

mandate. In the past, the election of the members has been delegated by the Convention to the ordinary (general) administrative conferences. The first elections were held by the Atlantic City Radio Conference, but, because there was no time to solicit nominations of individuals, countries were elected and were asked to appoint qualified experts. That conference determined that there would be eleven members of the Board at least until the next radio conference. The Buenos-Aires Plenipotentiary of 1952 established the number of members at eleven. The first and only regular election took place at the 1959 Administrative Radio Conference. The names of individuals and not countries were submitted and voted on.

After reducing the number of members of the Board from eleven to five, and in view of the fact that no general administrative conference was in the offing, the Montreux Plenipotentiary proceeded by way of exception to hold the third election, but at the same time confirmed that, in the ordinary course of events, the responsibility of holding elections was vested in the ordinary Administrative Radio Conferences. In view of the long interval between general administrative conferences — the last had been held in 1959 and the next was scheduled for 1979 — the Malaga-Torremolinos Plenipotentiary shifted the burden of election to the plenipotentiary conference which, at least in principle, was to meet at fairly frequent intervals. The delegates to the Malaga-Torremolinos Plenipotentiary did not feel it was appropriate for them to call an election without advance notice so decided, again as an exception, to turn the next election over to the Maritime Administrative Radio Conference which was scheduled for April, 1974, in Geneva.

The actual method of election that has been used is to request that administrations submit the names of qualified candidates in advance of the conference. A date is set aside during the conference for the actual election. The names of candidates are grouped according to the geographical regions they are from (Region A, the Americas; Region B, Western Europe; Region C, Eastern Europe; Region D, Africa; and Region E, Asia and Australasia) and all delegations are given the right to vote for one candidate in each regional grouping. The results of the IFRB election held in the World Maritime Administrative Radio Conference on April 29, 1974, were as follows:[18]

---

[18] See U.S., Department of State, Office of Telecommunications, *Report of the United States Delegation to the World Maritime Administrative Radio Conference of the International Telecommunication Union, Geneva, Switzerland, April 22 - June 7, 1974,* TD Serial No. 50, Washington, D.C., 1974 (mimeo), pp. 40-41.

Region A (The Americas):   Votes

Francis Gerard Perrin (Canada)   56(winner)

Fioravanti Dellamula (Argentina)   39

Blank Ballots   1

Region B. (Western Europe):

Charles William Sowton (United Kingdom)   47(winner)

Jose Maria Pardo Horno (Spain)   36

Dr. Carlo Terzani (Italy)   13

Blank Ballots   0

Region C (Eastern Europe):

Aleksandr N. Gromov (U.S.S.R.)   78(winner)[19]

Blank Ballots   18

Region D (Africa):

Abderrazak Berrada (Morocco)   69(winner)

Taofiqui Bouraima (Dahomey)   25

Blank Ballots   2

Region E (Asia and Australasia):

Sakae Fujiki (Japan)   65(winner)

Mihir Kimar Basu (India)   30

Blank Ballots   1

---

[19] Replaced in 1978 by Petr Sergeevich Kurakov, also from the U.S.S.R.

It should be noted that the U.S.S.R. is the only country that has been represented on the IFRB since its creation in 1947. After the first two elections this has been due to an unwritten rule which specifies that this high position in the ITU is the "property" of the U.S.S.R., just as the Director of the CCIR is the "property" of the United States and the post of Director of the CCITT is the "property" of France.

If, in the interval between two plenipotentiary conferences (ordinary radio conferences before 1973), a member of the Board "resigns or abandons his duties or dies," the Administrative Council normally replaces him at its next regular session, choosing from candidates proposed by countries in the same region. However, if the vacancy occurs more than ninety days before the regular Administrative Council meeting, then the country of which the former member was a national nominates a replacement, who will remain in office until the new member is elected by the Administrative Council.[20]

Organization and Methods of Procedure

The Convention provides that the IFRB elect its own chairman and vice chairman. The chairman serves for one year, after which he is succeeded by the vice chairman, and a new vice chairman is elected. Otherwise, the IFRB is given almost free reign.

In effect, the Board has organized itself as a collegial body. As such, all decisions are those of the Board as a whole and, in theory, must be defended by all of the members. Since it would be a practical impossibility for all of the members of such a body to have a complete knowledge of all of the aspects of the work involved, there is an allocation of responsibilities to the individual members based on preference, special knowledge, and the overall work load. The only area of continuity is the administration of the specialized secretariat which, by tradition, always falls on the shoulders of the chairman. Table 5.2 demonstrates how the Board recently distributed its responsibilities among the members.

---

[20] ITU, *International Telecommunication Convention, Malaga-Torremolinos, 1973,* Art. 57, para 2. (4).

The procedure is that the member in charge of a particular problem studies the problem with the assistance of a working group of the specialized secretariat and prepares a report on the matter for the Board. One or more informal meetings of the Board are held in which reports are presented and where suggestions are made and criticisms can be raised. If necessary, a question can be sent back to the member for further study. When agreement has been reached, or where no criticisms are raised, the reports are presented to the full Board at its weekly Friday morning meeting for a group decision. Since the matter has already been studied in some depth by the specialists in the board's secretariat and the member in question, and since problems have been ironed out in informal meetings, the work of the official Friday meetings is usually pro forma. The member in charge of the question makes his report and it is approved by the group as a whole without dissent. Three members constitute a quorum at the weekly formal meetings. These meetings are public and a record is kept of all of the Board's official actions in the working languages of the Union (English, French, and Spanish) and is available for public inspection at the offices of the Board.

This procedure is defended by the Board on the basis that it "has the merit of preventing discussions between the Board *in corpore* and the Chiefs of Departments, and thus avoids the raising of many problems of any kind which in no way concern the performance of the tasks of the Board."[21]

Rarely have the individual members of the Board expressed their disagreement in public. When it has occurred, it has not involved the substantive work of the board, but has involved primarily procedural matters and sometimes personality conflicts.[22]

The collegial atmosphere of the IFRB is reinforced by the Radio Regulations, which require that all formal decision be made by unanimous agreement. However, "If the Board fails in that endeavor, it shall thereafter decide the problem on the basis of a two-thirds

---

[21] See ITU, Administrative Council, 24th Session, Geneva, 3-23 May 1969, Doc. No. 3859-E (CA24-27), Annex, p. 2.

[22] For a discussion of one such incident, see David M. Leive, *International Telecommunications and International Law*, p. 79.

majority vote of the members present and voting for or against."[23] There has been only one recent occasion on which it was necessary to use the two-thirds majority alternative and that dealt with a personnel matter and not with the substance of the work of the Board. At one point a certain amount of subterfuge was necessary to keep the board's voting record clean. Since the U.S.S.R. did not sign the results of the 1951 Extraordinary Administrative Conference, its Board member felt that he could not approve of any actions carried out by the Board to implement those decisions. Rather than cast a negative vote, however, during the period when the 1951 decisions were an issue (from approximately 1952 to 1957), the Russian member made his position known and simply refrained from voting.[24]

Relations with Other ITU Organs

The IFRB is directly accountable to the Plenipotentiary Conference and the Administrative Radio Conference. The Plenipotentiary, through the convention, determines its structure and its general areas of competence. The radio conferences, such as the recent WARC 79, specify its organization and set forth its specific duties.

The relationship of the IFRB to the third representative body, the Administrative Council, is less clear. On the one hand, as the agent of the Plenipotentiary Conference, the Council is given a wide range of powers over the administration of the Union, including the approval of the annual budget. It also has a mandate to "review and coordinate the work programmes as well as their progress and the working arrangements, including the meeting schedules, of the permanent organs of the Union and take such action as it deems appropriate."[25]

---

[23] ITU, *Radio Regulations (Edition of 1976)*, Art. 11, 3 (4). This article also provides that "A quorum of the Board shall be one-half of the number of members of the Board. If, however, the verdict of such a quorum on a question coming before it is not unanimous, the question shall be referred for decision at a later meeting at which at least two-thirds of the total number of members of the Board are present. If these calculations result in a fraction, the fraction shall be rounded up to a whole number."

[24] From an interview with a former member of the Board on July 16, 1981.

[25] ITU, *International Telecommunication Convention, Malage-Torremolinos, 1973,* Art. 55, Para 10.j.

On the other hand, there is a certain ambiguity in the provisions of the ITU Convention in the articles setting forth the duties of the IFRB and the relationship between the two bodies. In particular, Article 10 states that one of the essential duties of the IFRB is to: "perform any additional duties, concerned with the assignment and utilization of frequencies and with the utilization of the geostationary orbit, in accordance with the procedures provided for in the Radio Regulations, and as prescribed by a competent conference of the Union, or by the Administrative Council *with the consent of a majority of the Members of the Union,* in preparation for or in pursuance of the decisions of such a conference."[26]

Along these lines, it is interesting to note that the IFRB does not submit its annual report to the Administrative Council for approval but rather sends it to the Council for "information purposes only" and that many of the individuals involved in the IFRB's work tend to give the impression that they are accountable only to the radio conferences and the Plenipotentiary and not to the Administrative Council.

Since the Board works in the same field as the CCIR, conflict has arisen at times over responsibility for interpreting the Radio Regulations. It seems clear, however, that both have such a right.

Since as early as 1947 an important problem has been the relationship of the IFRB with the ITU's General Secretariat. The IFRB, as we have noted, has its own independent functions given to it by the convention and the Radio Regulations and its members are elected by the Plenipotentiary Conference, as is the case for the Secretary-General. The effectiveness of the IFRB in the area of international frequency management, however, depends upon the size and competence of its secretariat. Although designated a "specialized secretariat" and thus under the control of the IFRB in the carrying out of its duties, it is also under the authority of the Secretary-General for administrative and personnel matters. Further, any additions to the specialized secretariat or any filling of vacancies must be approved by the Secretary-General and fit into the overall ITU budget for which the Secretary-General is responsible. Despite

---

[26] Ibid., Art. 20, para 3,d. Emphasis added.

the creation of the Coordination Committee, made up of representatives of all the four permanent organs, the matter of personnel has been a definite point of friction between the Secretary-General and the IFRB.

Conclusions

The IFRB is made up of skilled individuals who perform an important function in the maintenance of an efficient world telecommunications network.

The IFRB is not the quasi-judicial body that some of its originators thought it could be, nor is it the final arbitrator of interference cases between administrations that it was expected to be. It is primarily a recorder of frequency assignments for member states and, on occasion, aids in finding alternate frequencies when there is possibility of harmful interference.

On a number of occasions, such as the 1965 Plenipotentiary, it has been suggested that the job of recorder does not justify the existence of such a complex organ within the ITU. But, for a number of reasons, it has not been possible to shift the function to another organ such as the General Secretariat and thus divert the funds expended on the IFRB superstructure to other tasks. One reason is that a number of individuals in both the developed and developing countries see themselves someday in the rewarding position of member of the Board. A much more important reason is that the developing countries see the IFRB as an independent, non-aligned body which makes certain that their frequency notifications receive the same treatment as those of the developed states and, under certain circumstances, might even be persuaded to take the side of the developing country in a dispute with the administration of a developed country. Whatever the reason, it appears that the IFRB or its equivalent will be around for a long time to come.

TABLE 5.1
MEMBERS OF THE I.F.R.B., 1947 TO 1980

Name of Member	Country	Term Began	Term Ended
S. Banerji	India	1947	1952
A. H. Cata	Cuba	1947	1967
I. Danilenko	U.S.S.R.	1947	1950
F. Dellamula	Argentina	1947	1975
J. A. Gracie	United Kingdom	1947	1967
P. D. Miles	U.S.A.	1947	1953
R. Petit	France	1947	1975
N. H. Roberts	South Africa	1947	1967
J. J. Svoboda	Czechoslovakia	1947	1960
T. K. Wang	China	1947	1967
S. H. Witt	Australia	1947	1957
N.I. Krasnosselski	U.S.S.R.	1950	1956
S.S. Moorthy Rao	India	1952	1955
John H. Gayer	U.S.A.	1953	1966
P.S.M. Sundaram	India	1955	1960
B. Iastrebov	U.S.S.R.	1956	1960
R.E. Page	Australia	1957	1960
S. Hase	Japan	1960	1966
N.I. Krasnosselski	U.S.S.R.	1960	1965
M.N. Mirza	Pakistan	1960	1967
J. Ziolkowski	Poland	1960	1967
I.J. Petrov	U.S.S.R.	1965	1967
A. Berrada	Morocco	1967	
T. Nishizaki	Japan	1967	1975
V. Savantchuk	U.S.S.R.	1967	1971
A. Gromov	U.S.S.R.	1971	1978
S. Fujiki	Japan	1975	
F.G. Perrin	Canada	1975	
C.W. Sowton	United Kingdom	1975	
P. Kurakov	U.S.S.R.	1978	

TABLE 5.2

DISTRIBUTION OF TASKS BETWEEN
THE MEMBERS OF THE BOARD*

Task	Members Responsible
1. Technical examinations, findings and related recommendations	All Members of the Board
2. Notification procedures, technical examinations, technical standards, interference cases, special assistance, monitoring, preparation of related conferences, etc.:	
a. Space radiocommunications and radio astronomy	Messrs. Berrada, Sowton, Fujiki, Perrin, Gromov
b. Broadcasting service in its exclusive bands	Berrada, Fujiki, Souton, Gromov, Perrin
c. Services above 28 MHz: excluding space radiocommunications and radio astronomy and, in their exclusive bands, broadcasting, aeronautical mobile and maritime mobile services	Sowton, Fujiki
d. Services below 28 MHz: excluding broadcasting, aeronautical mobile and maritime mobile services in their exclusive bands	Gromov, Fujiki
e. Land mobile service	Sowton, Gromov
f. Maritime mobile service in its exclusive bands	Perrin, Berrada
g. Aeronautical mobile service in its exclusive bands	Perrin, Gromov
3. Technical standards: Series A, B and C	Fujiki, Gromov
4. Technical cooperation matters. Liaison with the General Secretariat with respect to all the questions related to the work of the I.T.U. experts, their reports and the establishment of technical recommendations and directives	Gromov, Fujiki
5. Coordination of the studies foreseen for the 1979 Conference	Berrada
6. Administrative, budgetary and and personnel questions	The Board through the Chairman

*From ITU, IFRB, Document D 14389 (Provisional edition), 4 July 1975.

chapter

THE ADMINISTRATIVE COUNCIL

The third non-permanent organ of the International Telecommunication Union is the Administrative Council, which meets in Geneva each spring for three weeks of deliberations. Composed of delegates from thirty-six member countries, elected by the Plenipotentiary Conference, the Council is the official surrogate of the Plenipotentiary Conference. However, for a number of reasons, including the fact that it meets for only a few weeks a year and that it represents only a small minority of the ITU's membership, it is only a pale reflection of its parent. While the Council carries out a number of important tasks for the Union, its influence on the other organs of the ITU has been limited and its impact on major policy issues has not always been significant.

Background

Prior to the 1947 Atlantic City Plenipotentiary, the ITU had no need for an Administrative Council. The plenipotentiary conferences, meeting at odd intervals, aided by world administrative conferences, were able to provide the necessary overall direction to the Union's activities; support and informational activities were carried out by the International Bureau and its Swiss staff. The Swiss Government managed the Bureau, appointing the Director and his staff, organizing it to carry out its mandate, controlling its finances, and auditing its accounts.

A number of factors, both internal and external, came together after World War II and changed the situation. The United Nations was in the process of being created by the victorious powers. Those existing international organizations which were to be permitted to remain outside the new body were at the least to be brought into a close relationship to the UN and its peace-preserving functions. It was felt that it would be inappropriate for the UN to enter into relationships with a secretariat, especially one with such a close relationship with one country, and the hard-to-work-with plenipotentiaries which were convened at such long intervals. A new organ was needed to provide more continuity and to speak for the ITU's membership.

Just prior to and during World War II the world had seen a tremendous increase in the use of radio; this would make new demands on the ITU, including the necessity of a new frequency list engineered to the needs of all nations of the world and a new international agency within the ITU to administer it. This new body would also need more supervision than the sporadic plenipotentiary conferences could exercise and more international guidance than could be supplied by the Swiss government under the existing Bureau arrangements. Finally, there was a feeling on the part of many that, especially in view of the anticipated increase in the tasks that the ITU was to perform, it would be inappropriate for the main service organ of the Union to remain under the control of a single state.

All of these factors came together in 1947 and, as a result, with minimum discussion and no criticism of the past performance of the Swiss government, it was decided to create an Administrative Council made up of delegates from member states to meet once a year to keep an eye on the work of the organs of the ITU and to supervise the Secretariat.[1] That body was to be composed of representatives of eighteen member countries elected by the plenipotentiary conference "with due regard to the need for equitable representation of all parts of the World" to supervise the new Secretariat and the finances of the Union, to arrange for the convening of ITU conferences, and in general to "perform any duties assigned to it by the plenipotentiary conferences."[2] In view of some of the recent criticisms that the Administrative Council does not provide enough continuity, it is interesting to note that the greatest amount of controversy at Atlantic City occurred over an unsuccessful Ameri-

---

[1]See Codding, *International Telecommunication Union,* pp. 201 & 286-291.
[2]See, *International Telecommunication Convention, Atlantic City, 1947. Art. 5.*

can proposal that the Council have a permanent executive committee which would sit between Council sessions to assure that the work of the Union would be supervised on a regular basis.

Almost every plenipotentiary has made changes in either the structure or the functions of the Administrative Council. The most important structural changes have been the constant increases in the number of members of the Council which have resulted directly from the increase in the number of members of the Union, and the reluctance of countries from any region of the world represented on the Council to have their share reduced. Starting with eighteen members in 1947, the Geneva Plenipotentiary of 1959 raised it to twenty-five, the 1965 Montreux Plenipotentiary to twenty-nine, and the Malaga-Torremolinos Plenipotentiary of 1973 to its present thirty-six.

Functions

As the official agent of the plenipotentiary conference, the Administrative Council is responsible for the proper administration of all aspects of the International Telecommunication Union between plenipotentiaries. The International Telecommunication Convention, which is the handiwork of the plenipotentiary conference, attempts to spell out the exact meaning of this relationship. The Basic Provisions of the Convention articulate three major areas of concern for the council: facilitating the implementation of the convention, regulations, and decisions of the various ITU conferences by member countries; ensuring the efficient coordination of the work of the Union and exercising financial control over its organs; and promoting international assistance for the provision of technical cooperation to the developing countries.[3]

The General Regulations section of the Convention, which goes into greater detail, can be compacted into three major categories: external relations, supervisory activities, and administration. External relations are the least time-consuming of the duties of the Council, and involve formal contacts between the ITU and the United Nations, the Specialized Agencies of the United Nations, and other

---

[3]*International Telecommunication Convention, Malaga - Torremolinos, 1973*, Art. 8. Art. 8.

intergovernmental agencies, including the negotiation of provisional agreements with them. The supervisory duties, which are extensive, include the drawing up of such regulations as it considers necessary for the administrative and financial activities of the Union, supervising the administrative functions of the Union, reviewing and approving the annual budget of the Union, arranging for the auditing of the accounts of the Union, and reviewing and coordinating the work programs of the permanent organs of the Union. Also, the Council makes arrangements for the holding of conferences, when necessary, and, if it is an administrative conference, establishes its agenda in consultation with member administrations. Among the specific administrative duties of the Council are the filling of vacancies in the posts of Secretary-General, Deputy Secretary-General, Director of the CCIR, Director of the CCITT, and members of the IFRB if vacancies occur between meetings of the plenipotentiary. The tasks dealing with secretariat personnel matters are of a much more time-consuming nature. The Council is responsible for staff regulations and rules, the number and grading of the staff, and the adjustment of salaries and allowances as necessary within the United Nations Common System. The Administrative Council also draws up a report of its own activities, which is sent to each of the member countries.

For years the Administrative Council labored under the realization that although it represented the plenipotentiary, it contained only a small majority of the total membership of the Union; consequently it would be criticized if it went too far, just as it would be criticized if it did not go far enough. This problem was solved to some extent in 1959, when the Geneva Plenipotentiary Conference of that year gave the Council blanket authority to "take the necessary steps . . . provisionally to resolve questions which are not covered by the Convention and its Annexes and cannot await the next competent conference for settlement" as long as the Council obtains "the agreement of the majority of the members of the Union."[4] When in doubt, or if a serious conflict should occur, the Council simply puts the issue to a vote of the members by correspondence.

---

[4]*International Telecommunication Convention, Geneva, 1959*, Art 9, 18, 0. Now contained in *International Telecommunication Convention, Malaga-Torremolinos, 1973*, Art. 55, 10, 0.

Composition

The Administrative Council is made up of representatives of thirty-six member countries elected by the plenipotentiary conference "with due regard to the need for equitable distribution of the seats on the Council among all regions of the world."[5] Countries hold office for the interval between plenipotentiaries. If a seat becomes vacant between meetings of the plenipotentiary conference, it passes to the member of the Union from the same region which obtained the next largest number of votes among those not elected.[6]

The present membership of the Administrative Council was elected at the Malaga-Torremolinos Plenipotentiary Conference on October 3, 1973. The first step was a call for the submission of names of countries interested in being represented on the Council grouped according to the five geographical areas first used at the 1959 Geneva Plenipotentiary: Region A (The Americas), Region B (Western Europe), Region C (Eastern Europe and Northern Asia), Region D (Africa), and Region E (Asia and Australasia).[7] Delegations at the plenipotentiary were given ballots which listed all nominations grouped by region. Delegates were permitted to vote for the designated number of countries in each of the five regions; seven seats in Region A and B, four in Region C, and nine in Regions D and E. There was no choice for Regions B and C, inasmuch as the delegations from those regions met prior to the voting in plenary session and narrowed the number of nominees to the exact number of seats to be filled. Region A had eleven candidates, Region D eighteen, and Region E fourteen. The countries elected to send delegates to the new Administrative Council were:[8]

---

[5] *International Telecommunication•Convention, Malaga-Torremolinos, 1973*, Art. 8,i(1).

[6] A seat is declared vacant if the member involved resigns its membership on the Council or if it does not have a representative in attendance at two consecutive annual sessions of the Council.

[7] Prior to the 1959 Plenipotentiary there were only four regions. A, B, and C were the same but the D group contained all "other countries."

[8] See ITU, *Minutes of the Plenipotentiary Conference of the International Telecommunication Union, Malaga-Torremolinos, 1973*, Geneva, 1974, pp. 292-298.

Region A (The Americas)	votes	**Region D (Africa)**	
1. Mexico	103	1. Tanzania	86
2. Canada	103	2. Egypt	80
3. U.S.A.	100	3. Algeria	76
4. Brazil	96	4. Morocco	75
5. Argentina	89	5. Cameroon	73
6. Trinidad and Tobago	67	6. Senegal	70
7. Venezuela	67	7. Nigeria	69
		8. Zaire	68
Region B (Western Europe)		9. Ethiopia	66

Region B (Western Europe)		**Region E (Asia and Australasia)**	
1. France	118		
2. Germany (F.R.)	115		
3. Spain	114	1. Japan	114
4. Italy	113	2. India	101
5. Switzerland	111	3. China	97
6. United Kingdom	108	4. Australia	94
7. Sweden	103	5. Saudi Arabia	94
		6. Lebanon	86
Region C (Eastern Europe and		7. Thailand	74
Northern Asia)		8. Iran	72
		9. Malaysia	63
1. Roumania	116		
2. Poland	114		
3. Hungary	110		
4. U.S.S.R.	108		

The following countries have been represented on the Administrative Council since 1947: Canada, U.S.A., Brazil, Argentina, France, Italy, Switzerland, U.K., U.S.S.R., and China. Mexico, India, Poland, and Egypt have all been elected four times.

The Convention provides that each country elected to the Council shall appoint one person to serve on the Council "who may be assisted by one or more advisers."[9] At least since the 1973 election, the favorite size of delegations to the Council has been two or three, although each time a few member countries have sent more and a few less. In 1980, for example, twelve members sent four or more

---

[9]*International Telecommunication Convention, Malaga-Torremolinos, 1973*, Art. 8, 1. (2).

representatives, including Japan with seven; France, Sweden, and the U.S.S.R. with six; and Germany (F.R.), United Kingdom, and the United States with five. Six countries in 1980 sent only one: Ethiopia, Malaysia, Mexico, Morocco, Trinidad and Tobago, and Zaire.[10] The Convention also provides that: "The person appointed to serve on the Council by a Member of the Administrative Council shall, so far as possible, be an official serving in, or directly responsible to, or for, their telecommunications administration and qualified in the field of telecommunication services."[11] As far as can be determined from the information presented by the ITU in the annual list of participants, as a general rule, the countries elected to the Council send as their chief representative an individual who fulfills at least the first of the stated qualifications. Almost all of the chief delegates to the 1979 council meeting, for instance, were in the top administrative echelon of their respective telecommunications or PTT ministries, although seldom the highest administrator, and the favorite position seems to be the director of external relations. If the chief delegate is an administrator rather than an individual qualified in the technology of telecommunications, quite often one or more such individuals are to be found in the remainder of the delegation.[12]

While expertise and authority are important elements in the make-up of delegations to the Council, continuity can also be an important factor in how well a delegation carries out its responsibilities. Of the delegations to the 1980 council meeting, eleven delegations could boast of having an individual in their group who had seen service since the Council was elected in 1973; Australia, Cameroon, Canada, Germany (F.R.), Hungary, Mexico, Poland, Roumania, Senegal, Spain, and Sweden. Six had individuals who had attended six out of the seven sessions: China, France, Lebanon, Tanzania, Thailand, and the U.S.S.R.. There were also a number of countries with delegates who had attended four or five of the seven sessions. Only four delegations contained delegates with no previous experience: Ethiopia, Iran, Malaysia, and Zaire.

---

[10]See "35th Session of the ITU Administrative Council," *Telecommunication Journal*, Vol. 47, No. 8 (August 1980), pp. 494-495.

[11]*International Telecommunication Convention, Malaga-Torremolinos, 1973*, Art. 55, 2.

[12]See ITU, Conseil d'Administration (34e Session), *Liste des Participants*, Geneva, June 4, 1979 (mimeo.).

Only the travel and subsistence expenses of the chief representative of a country to the Council are reimbursed from the ITU's general budget. Salaries of councillors remain the province of the sending administration.

The Council at Work

This section of the chapter will deal with three interrelated subjects, the general rules governing the work of the Council, a case study of the Council at work, and a discussion of some of the major problems that the Council has encountered in recent years.

The official ITU documents contain only a few rules concerning the work of the Council. For instance, the Council must hold annual sessions at the seat of the Union. Additional meetings can be held at the will of the Council or may be convened by its chairman at the request of a majority of its members. As a general rule, the Secretary-General, Deputy Secretary-General, Chairman, and Vice-Chairman of the International Frequency Registration Board, the Director of the CCIR, and the Director of the CCITT have the right to participate in the work of the Council but not the right to vote. However, if it so decides, the Council can hold meetings confined to its own membership. The Secretary-General of the ITU acts as the secretary of the Administrative Council.[13]

The Administrative Council elects its chairman and vice-chairman at the beginning of each session and they serve until the opening of the next session. In practice, unless something untoward should occur such as his country not being re-elected to the Council, the vice-chairman automatically succeeds to the chair. In distribution of representation, these two offices do not tend to be monopolized by any one segment of the Council. The following is a list of the chairmen of the Administrative Council over the last ten years:

1981 — Gheorghe Airinei (Roumania)
1980 — Masakatsu Yonezawa (Japan)
1979 — Jean Jipguep (Cameroon)
1978 — Heinrich L. Venhaus (Germany, F.R.)
1977 — Joao Santelli (Brazil)
1976 — Laszlo Katona Kis (Hungary)

---

[13] *International Telecommunication Convention, Malaga-Torremolinos, 1973* Art. 55.

1975 — Maurice Ghazal (Lebanon)
1974 — Victor A. Haffner (Nigeria)
1973 — Aurelio Ponsiglione (Italy)
1972 — Aldo Santiago Irrera (Argentina)
1971 — Ahmed Zaidan (Saudi Arabia)

The Rules of Procedure adopted by the Council do not differ greatly from the rules used by other international organizations as concerns accreditation of members, duties and powers of the chairman, conduct of debates, establishment of committees and working groups, records of meetings, and the like. The rules do, however, provide for a reminder by the Secretary-General two months in advance of a meeting and the sending out of an Agenda for the next meeting to members of the Council and the Secretary-General of the United Nations as soon as possible after the Council adjourns. All such agendas include consideration of the annual report of the activities of the Union, the draft annual budget, and other items submitted by the Secretary-General of the ITU, members of the Union, conferences or permanent ITU organs, or by the United Nations or a Specialized Agency of the United Nations. The United Nations is entitled to send an observer to council meetings (without the right to vote) as can any Specialized Agency of the United Nations which has been invited by the Council. Proposals to the council involving expenditures must be accompanied by a separate statement by the Secretary-General of the actual costs involved.

Voting procedures include the exhortation that the Council attempt to achieve a consensus "so that it is unnecessary to take a vote," but if consensus is not reached, voting may take place by (1) a show of hands, (2) a roll call vote, and (3) a secret ballot. A simple majority is all that is needed, but a proposal is not adopted unless 24 members with the right to vote are represented and the proposal has the affirmative vote of at least 16 members. In the case of a tie, the measure is considered to be rejected; if the number of abstentions exceeds one-half of the number of members voting, the measure must be reconsidered at a later meeting, at which time abstentions are not taken into account. Finally, a decision may not be reconsidered during the same session unless a majority of the Council decides otherwise.[14]

---

[14]See ITU, *Rules of Procedure of the Administrative Council of the International Telecommunication Union (1976 Revision)*, Geneva, 1976.

Before proceeding, one additional Council procedure merits attention. In the early post-World War II years, there was no limit on the number of committees. At its 1962 meeting, for instance, the Council established committees for finances, staff and pensions, relations with the United Nations and other international organizations, audit of accounts, and drafting. It also established two working parties, one for outer space communications and one on ITU centennial year preparations. In more recent years, the Council has restricted its formal structure to three major committees consisting of: Committee 1, Finance; Committee 2, Staff and Pensions; and Committee 3, Technical Cooperation. All committees are committees of the whole.

In the past it was also the custom for the chairs of committees to be held by the same country for long periods of time. In the early days of the Council, for instance, the committee dealing with relations with the United Nations and other international organizations was almost the personal fiefdom of the American, Francis Colt de Wolf. India was elected time and again as chairman of the Finance Committee, although the individual involved changed more often than his U.S. counterpart, and before he became Secretary-General of the ITU, Mohamed Mili held the chair of the Committee on Audit of Accounts for several years. In more recent times, the Committee on Development Assistance was chaired for a long period by an American, the Committee on Staff and Pensions by the United Kingdom, and the Committee on Finances by Switzerland.[15] In 1977 Switzerland gave up the chairmanship of the Finance Committee and in 1978, with the retirement of Tom Ulrik Meyer, the Staff and Pensions Committee passed out of the hands of the United Kingdom. In each case, a representative of a developing nation took over. In the case of the United Kingdom, it was Morocco; and in the case of Switzerland, it was Mexico, followed by India.

At its 1979 meeting, under the chairmanship of Jean Jipguep of Cameroon, the Council made a number of decisions that tended to assure that the new system would be continued. It was agreed in

[15]The American, Thomas E. Nelson, for instance, held the chairmanship of Committee 3 from 1968 to 1975, when it was taken over by John O'Neill, Jr. O'Neill retired from the U.S. Foreign Service in 1978 when the chairmanship was taken over by Mrs. Ruth H. Phillips, also an American. An Englishman, Charles E. Lovell, held the chair of Committee 2 from 1969 to 1973 when it was taken over by Tom Ulrik Meyer of the U.K. until his retirement in 1977. For many years Rudolf Rutschi was the Swiss chairman of the Finance Committee.

principle that chairmanships should rotate in the future. Individuals rather than countries would be appointed to those positions while preserving the concept of equitable geographical distribution. Councillors would serve for a minimum of two years and a maximum of five.[16] In 1980 the United States lost to Senegal the chairmanship of the Technical Cooperation Committee.

The 1980 session will be used as an example to show the actual day-to-day work of the Council. Thirty-two delegates and their sixty-two assistants and aids arrived in Geneva over the weekend of May 10. Each delegate had a draft agenda prepared by the Secretary-General, along with a report on each of the items to be considered. On the morning of Monday, May 12, the councillors and aids congregated at the entrance to the conference room in the ITU tower where they were checked in and were given lapel passes to permit access to the conference room itself. As soon as the preliminaries were completed, the meeting was called to order by the previous chairman, Jean Jipguep of the Cameroon. The Council then proceeded to elect Masakatsu Yonezawa, who had served as vice-chairman at the 34th Session, as the new chairman. After thanking his colleagues for the honor to him and his country, the new chairman led the council through the routine of electing a vice-chairman (Gheorghe Airinei of Roumania) and the chairmen and vice-chairmen of the three standing committees. The word "routine" is used advisedly inasmuch as the names of the individuals who were elected had been chosen prior to the meeting through a number of consultations between Jipguep, the Secretary-General, and individuals in member administrations. The chairmen and vice-chairmen elected by the first plenary meeting were as follows:

Committee 1 — Finance
   Chairman: T.V. Srirangan (India)
   Vice Chairman: N.J. Mazzaro (Argentina)

Committee 2 — Staff and Pensions
   Chairman: Abdeljalil Zrikem (Morocco)
   Vice Chairman: J.-P. Duplan (France)

Committee 3 — Technical Cooperation
   Chairman: M. Samoura (Senegal)
   Vice Chairman: L. Katona Kis (Hungary)

---

[16]See U.S. Department of State, Report of the U.S. Mission in Geneva to the Secretary of State, International Telecommunication Union: 34th Session of the Administrative Council, Incoming Telegram, No. 10564.

The Council then turned to the important problem of the agenda. The final agenda that was adopted by the First Plenary Meeting of the 1980 Administrative Council consisted of a total of sixty-one items. Twenty-six items were reserved for consideration in plenary meetings, sixteen were given to Committee 1, ten to Committee 2, and nine to Committee 3. The 35th Session of the ITU Administrative Council also had the assistance of two working groups. One, the Working Group on the Future of ITU Technical Cooperation, was a carry-over from the 34th Session of the Council. The other, the Working Group PL/C, was established by the 35th Session to consider the implications of the decisions of the 1979 World Administrative Radio Conference on the work of the permanent organs of the Union and a report on the additional computer needs of the International Frequency Registration Board.

The ITU Council probably spends more time in plenary session than most comparable bodies because of the large number of issues that it feels demand its attention. A number are fairly routine, such as the ITU's involvement in the Transport and Communications Decade in Africa and in the World Communications Year. Some are important but take little time, such as the approval of the annual report to member administrations which has been drawn up by the Secretariat, the date and duration of the next Council meeting, and changing the rules dealing with the ITU's Centenary Prize. The council also tends to take up other matters in plenary meetings that seem to be committee matters, such as the use of computers by the IFRB and the CCI's publication policies, changes to the premises of the ITU, and the like.

The Committee on Finances seems to carry the heaviest work load of the three, followed by the Committee on Staff and Pensions, which involves itself quite intimately with personnel problems of the Secretariat. Despite its subject matter, the Technical Cooperation Committee might be considered the least hard-working of the three Council committees if it were not for the fact that recently it has been involved in establishing guidelines and exploring the parameters of the ITU's involvement in technical assistance activities. It is, in effect, one of the first attempts on the part of the Administrative Council to seriously get into the realm of long-term planning and decision making in an important aspect of the functions of the ITU.

In the day to day activity of the Council, the early part of the session is interspersed with plenary assembly sessions and committees. As the committees make their reports, the number of plenaries increases up to the final days when only plenary sessions are held to make final decisions on the agenda items. One of the last such decisions is the date of the next meeting. After the chairman and the vice-chairmen make their final speeches, the delegates rush off to their respective homes to take up their ordinary duties. As a general rule, the council meetings take up about nineteen full work days.

Reform

The Administrative Council is not unaware of the possibility that it may not be providing the ITU with the type of leadership that it needs. Innumerable sessions have been devoted to discussion of ways to improve its performance and, on two occasions, it has entered into the matter in some depth. The first occurred in meetings between 1948 and 1965 when the Council investigated methods of providing for better coordination between the heads of the permanent organs on decisions affecting them all, especially those dealing with personnel, in order to relieve the Council of some of its burden. This investigation resulted in the creation of the Coordination Committee and its elevation to the position of a quasi-permanent organ of the ITU. This committee and its contributions will be discussed later.[17]

At this point, we will look at the Council's investigation of working methods during the period 1977-1979. The first discussion took place at the 1977 session. A number of the councillors at that time were of the opinion that if the Council were more efficient, it would free up some time for the consideration of more important, long-term issues. As has often been the case, it was found that time did not permit a full discussion of the subject. Rather than postponing the issue to the next meeting, however, the Council decided to establish an intersessional working group to study the issue and to draw up a report for the 1978 Council on possible improvements in Council working methods.[18] A meeting of the working group was held in Madrid from

---

[17]See below, pp. 174-178.
[18]Members of the group were Algeria, Canada, France, Germany (F.R.), Hungary, Italy, Japan, Morocco, Spain, Sweden, U.K., U.S.A., and U.S.S.R.

the 5th to the 7th of April, 1978, to discuss suggestions that had been submitted from Germany (F.R.), France, Canada, Hungary, U.K., and Sweden. The group first agreed that "it is essential to improve the Council's working methods, and especially to ensure that all the items on its agenda are given adequate consideration, to prevent the unduly frequent postponement of items to the next session, to ease the work of the Councillors and to render decisions clearer and more effective."[19] Three areas of special concern were then identified: (1) organization of work and preparation for meetings, (2) documentation, and (3) budgetary problems.

The possible solutions to the first area of concern that were debated but rejected because "no consensus was reached in favor of such a solution" included the holding of biennial meetings of the Council and extending the terms of the Council's Steering Committee. Those proposals which were approved, at least in principle, include the more extensive use of sessional working groups. It was also decided that such groups should not meet outside council sessions except in exceptional circumstances. The working group also suggested the possibility of (1) drafting a provisional agenda for the following session during each year's meeting to help councillors better prepare themselves to discuss it; (2) reducing the size of the Annual Report "by omitting additional information which was not of fundamental importance and which was already known to the Members"[20] in order to cut down the time the Council devoted to its consideration; and (3) having the Secretary-General issue quarterly reports on the activities of the Union to the members of the Council "that would enable them to follow the activities of the Union closely, in particular with regard to management, staff matters and finance."[21] Changes in the duration of Council meetings were discussed, but "They did not favor laying down a shorter period than nineteen days as a general rule."[22]

As regards documentation the working group proposed the adoption of a standard format and a longer lead time in receiving the documents. The group felt that the dispatch of documents four weeks

---

[19]See ITU, Administrative Council, 33rd Session, Geneva, May/June, 1978, Doc. No. 5238-E (CA33-80), 23 May 1978. p. 2.

[20]Ibid., p.3.

[21]Ibid.

[22]Ibid., p.4.

before the opening date of the Council was too short to allow adequate study, especially for those which had "significant financial, organizational, or staff implications." That type of document, it was agreed, should reach the councillors at least two months before the opening of the annual meeting.

To solve the Council's problems in giving the budget adequate consideration, the group suggested that a study be made of the practice of the Universal Postal Union and other international organizations to see how their budgets are presented, when in the time frame of the annual meeting they are discussed, and how much time is devoted to them. It was also agreed that the Secretary-General should, as a rule, present a budget within the limits set by the Convention. If, "inexceptional circumstances," he considers it unavoidable to present a budget in excess of the limit, "he should clearly identify, in an annex, the operations in excess of the limit, justify them and offer options which, if adopted, would bring the Budget within the limits."[23]

Finally, the working group voiced the pious sentiment that, at the annual meetings, "Councillors must discipline themselves to restrict their interventions in both length and number to avoid digression."[24]

The report of the working group was presented to the full Administrative Council at its 33rd Session in Geneva in 1978 and a lengthy debate ensued. The full Council approved three of the working group's suggestions. It was agreed that it would be useful to have a provisional agenda for the following session before the end of each session and the Secretary-General was requested on a trial basis to provide one before the end of the 33rd Session. Also, to facilitate the flow of information to the members of the Council, it was agreed that the Secretary-General should experiment with a quarterly newsletter in which he would endeavor to keep the members of the Council up to date with important developments. The newsletter should not deal with the routine activities of the organs of the Union, but only events of particular significance such as "the follow-up given to Council decisions and the status of important items that would come

---

[23]Ibid., p.5.
[24]Ibid., p.2.

up for discussion and decision at the next Council session."[25] For example, he should include consultations held with the Administrator of the United Nations Development Program, "developments with respect of the unified approach to technical cooperation within the United Nations Common System," information on the budgetary situation, "and any unusual measure which the Secretary-General might need to take — for instance, in the public relations field — during the course of the year."[26]

It is interesting to note that during the discussions on this item the perennial issue of the exact nature of the Council's authority over the activities of the Union's permanent organs, especially the Secretary-General, was broached. One councillor was emphatic in his opinion that the newsletter should not contain routine items concerning the management of the Union since that was the responsibility of the Secretary-General. The Council's function as opposed to that of the Secretary-General was primarily legislative. Another councillor objected to that point of view because he felt that there was no question but that the Administrative Council was responsible for the management of the Union, therefore it should have as much information on management as it needed to do its task in as effective a manner as possible. The third decision of the 33rd session of the Council was to approve the new standard format for Council documents that had been suggested to the working group by the representative of the Federal Republic of Germany.

Most other points raised by the working committee were either put aside or postponed for consideration at a later session. The entire section on working methods, for instance, was discussed but no action was taken. The question of the creation of permanent working parties was postponed, as well as the issue of amending the four week rule for the submission of documents by the Secretary-General to the councillors before each session. As regards the admonition by the working group to members of the council to exercise discretion when addressing each other in meetings, the intervention of the Canadian delegate during the 33rd session is worthy of reproduction in full:

---

[25]See ITU, Administrative Council, 33rd Session, Geneva, June 1979, Document No. 5260-E (CA 33-102), 8 August 1978, p. 6.

[26]Ibid. p.6.

> *"Mr. Warren* (Canada) said that special attention should be paid to the last two paragraphs of point 3, relating to the approach adopted by the Councillors and to the need for self-discipline. The current session was fortunate in being less pressed for time than others had been and in having four excellent Chairmen, so that there was less need than usual for the imposition of such disciplinary measures as limiting the length and number of interventions; yet he could not help noting that, for instance, it had been unnecessary for practically every Councillor to congratulate the new member of the IFRB on his election, since statements by representatives of regions would have sufficed."[27]

The Product

No major document exists concerning the results of the work of the meetings of the Administrative Council that embodies the real contribution of the Council to the ITU, as is the case of the Convention for the Plenipotentiary Conference or the Telegraph and Telephone Regulations for the World Administrative Telegraph and Telephone Conference, for instance.[28] The work of the Administrative Council consists mainly of a series of decisions, recommendations and suggestions directed to the various permanent organs of the ITU, especially the Secretariat, to help them in their day-to-day work and to keep them from transgressing the limits to their activities that have been established by conferences of representatives of members of the Union. As such, they are of a temporary nature and tend to be quickly outdated.

If one were forced, however, to choose one document that could be consulted to gain a general idea of the nature of the actual work that the Council performs, it would have to be the document entitled "Resolutions and Decisions of the Administrative Council of the International Telecommunication Union." This loose-leaf document is revised in its entirety by the Council every few years and, in

---

[27]Ibid., p.3.

[28]See below, Chap. 10.

the interim, brought up to date by annual additions. The latest such document was last revised in 1981. The bulk of the document consists of a collection of those resolutions and decisions that were considered to be still valid after the 1981 revision as augmented by the new decisions and resolutions made by the 36th Session in 1980, organized along the following categories:

Section 1. Finances
Section 2. Personnel, Pensions
Section 3. Conferences, Meetings
Section 4. Permanent Organs

Section 5. Radiocommunications
Section 6. Technical Cooperation
Section 7. External Relations (United Nations, etc.), Information
Section 8. (for future use)
Section 9. Miscellaneous

Much of the routine history of the ITU is contained in these documents, as well as a good indication of the value of the Administrative Council to the successful operation of the ITU over the years. Appendix A reveals that the Administrative Council issued 839 resolutions and decisions in the period from 1948 to 1979.

And the cost of the Council? If one ignores the salaries of the members of the Council and other emoluments that are paid by the sending administrations during the three weeks that the councillors are in Geneva, and the cost to administrations of travel to Geneva, which can be relatively high for the smaller and less wealthy nations, the entire expenditure of the Council in 1979, for instance, amounted to 708,931.80 Swiss francs.[29]

Conclusions

As one of the newest organs of the ITU, the Administrative Council has appropriated an important place for itself in the activities of the ITU. In addition to its role as the contact point between the ITU and other international organizations, it has taken over a great deal of

---

[29]See ITU, *Report of the Activities of the International Telecommunication Union in 1979*, Geneva, 1980, p. 130.

the supervision of the other organs of the ITU, especially the General Secretariat. While it is probably no more efficient in its task than was the Swiss Confederation in the years before 1948, it is also extremely doubtful that the ITU's membership would ever countenance a return to the old system.

The Administrative Council appears to have two interrelated problems: too much to do in the time at its disposal and a conflict as to the limits of its responsibility for the activities of the other organs. The amount of work that the Council undertakes is determined to a certain extent by the increase in size of the membership of the ITU and the Administrative Council itself, and in its functions, especially those dealing with the radio side of telecommunication. But the cause also lies in the scope of the responsibility it assumes over the work of other ITU bodies.

There are two obvious remedies, one organizational and one functional. Since there is no serious drive to cut down on the size of the council — a move which would give it greater flexibility and thus probably result in increased efficiency — the organizational solution to the problem of overwork would be to adopt a policy of creating more working groups to carry out more investigations between sessions, to extend the session of the Council to more than nineteen days, to have biennial sessions with separate agendas, to increase the power of the Council's Steering Committee, or to establish an executive committee of the Council with the power to make certain decisions in the interim between Council meetings. All these solutions have been discussed at one time or another since the Council was created in 1947, and all have received some degree of support among the members.

The solution would be to reduce the scope of the functions of the Council in order that it could assume more of a supervisory role over the activities of the ITU's organs rather than being an active participant in their day-to-day work. It is impossible for the Council to investigate basic issues such as how to make the Union more responsive to the needs of the new developing majority and still perform tasks important to the developed nations while it is wrestling with personnel matters or even the computer time needs of the IFRB and the CCIR.

Because of the limited time available at annual meetings, it is doubtful whether the Council is in a position to make any fundamental changes in its working methods. The proper organ to do so is probably the plenipotentiary conference, and an occasion to do so will arise in 1982 in Nairobi. However, in view of the many other issues that it must address, it is doubtful whether the 1982 Nairobi Plenipotentiary Conference will have the time necessary to give the Administrative Council anything more than the most cursory advice.

chapter 7

THE
SECRETARIAT

The General Secretariat

The foundation on which every international organization rests is its Secretariat. The Secretariat does not come and go as do delegates to conferences and meetings; it works day by day at the headquarters in Geneva. The Secretariat provides continuity and permanence. It is not surprising, therefore, that the Secretariat of the ITU is often mistaken for the organization itself.

The Secretariat performs functions without which an international organization could not exist. In the first place, it provides the support services necessary for conferences, the meetings of the Administrative Council, and meetings of the plenary assemblies and working groups of the Consultative Committees. Second, it is the distribution point for information of all types dealing with the work of the ITU and the international telecommunication network in general. And third, it is the daily contact point between the ITU and member administrations and between the ITU and other international organizations.

The purpose of this chapter is to portray the structure and functions of this essential component of the International Telecommunication Union and some of the problems associated with it.

Background

Along with the IFRB and the Administrative Council, the General Secretariat was the third major organizational innovation of the 1947 Atlantic City Telecommunication Conference. Prior to 1947, the functions of the Secretariat were essentially carried out by the

Swiss government. It was the Swiss government which provided the personnel of the Bureau, it was the Swiss government which nominated the individual who was to serve as the director of the Bureau, and it was the Swiss government that provided the funds for the operation of the Bureau and audited its accounts. The individuals nominated to the post of director by the Swiss government were always men of substance with a knowledge of telecommunications, although seldom innovators. Their loyalty, however, was first of all to Switzerland and not to the members of the ITU; the Swiss government was thus privy to inside information concerning problems or politics that surfaced in the small, ITU Bureau.

It was this situation that a number of delegations to the 1947 Atlantic City Plenipotentiary Conference were determined to change. The first influence in this direction was undoubtedly the successful experiment in an internationalized secretariat exemplified by the League of Nations and, more importantly, by the International Labor Organization. The new United Nations and its affiliated international organizations were to follow the same pattern. The idea of having personnel of various nations working together, combined with the concept of an "international civil service," had an appeal that was hard to resist. The delegates to Atlantic City were also uncomfortable with one country having the influence that Switzerland had over the activity of the Secretariat, especially in view of the new activities that the Atlantic City conferences were planning for the ITU. This feeling was reinforced by the fact that the victors of World War II were in charge of organizing the peace. Swiss neutrality was viewed askance by several of the victors, despite the fact that it had been the deciding factor in putting the Bureau in Swiss hands in the first place and, in effect, had ensured the continuity of the ITU in both the first and the second World Wars. In any event, the decision at Atlantic City was overwhelmingly to internationalize the Bureau and to provide for the election of its director by members of the Union. It did not, however, try to move the new secretariat out of Switzerland, but to change its location from the capital of Switzerland to Geneva, the old League and new U.N. European headquarters.

It is interesting to note that the delegates to the Atlantic City conference were in complete agreement that the new secretariat should have no independent powers whatsoever, especially powers of initiative. Not only did this consensus aid in putting down the fears of delegations that the secretariat would do something harmful to their collective "sovereignties," but it would also ensure the individuals who served in the secretariat of "permanence in their employment."[1]

The new ITU Secretariat was quickly put to the test. It was immediately necessary to begin the physical move of personnel, equipment and files from Berne to Geneva and to hire nationalities other than the Swiss to be a part of the new international Secretariat. All of which was tremendously complicated by the ITU engaging in an unprecedented number of conferences in a very short period of time. In addition, it was necessary to sort out the lines of authority over personnel between the new General Secretariat and the new IFRB and the three consultative committees, each of which had been given its specialized secretariat by the new Atlantic City Convention. That the new Secretariat was able to successfully carry out its move, its internationalization, and the tremendous burden of providing support for the numerous conferences that were held in the period 1948 to 1952, attests to the abilities and devotion of the officials of the renovated Bureau. Only the matter of lines of authority was not worked out and, as will be seen later, they exist in the ITU to this day.

Other than an increase in size, due to increased duties and an increase in members of the Union, the form of the General Secretariat of the ITU remains essentially as it was created in 1947.

Functions

The official functions of the General Secretariat, as set forth in Article 9 of the ITU Convention, are written in terms of the Secretary-General and not in terms of a secretariat: "he shall be responsible to the Administrative Council for all the administrative and financial aspects of the Union's activities."[2] Article 56 of the General Regulations section of the Convention contains the details. As with any collection of this kind, it contains reminders, suggestions, and directives, some of which are spelled out in great detail and others which are only reflections of the general sentiments of the delegates to the plenipotentiary who wrote them.

---

[1]See Codding, *The International Telecommunications Union*, p. 292.

[2]See ITU, *International Telecommunications Convention, Malaga-Torremolinos, 1973*, Art. 9, 1(3).

The Article 56 functions of the Secretary-General can be divided into four general categories: (1) general administration; (2) staff; (3) work in support of conferences and meetings; and (4) the dissemination of information.

On the administrative level, the Secretary-General of the ITU is charged with coordinating the work of the permanent organs with the assistance of the Coordination Committee (a subject that will be dealt with in some detail later in this chapter), organizing the work of the General Secretariat, ensuring the application of the financial and administratative regulations approved by the Administrative Council, preparing and submitting future work plans to the council, preparing cost-benefit analyses of the work of the headquarters for the Council when requested and providing legal advice to the organs of the ITU. The Secretary-General also submits an annual budget estimate to the Administrative Council, prepares an annual financial and operating report for the Administrative Council, and prepares an annual report of the activities of the Union, which, after approval by the Administrative Council, is transmitted to all of the members of the Union.

With regard to the staff of the ITU, the Secretary-General organizes the work of the General Secretariat, appoints its members and supervises it "with a view to assuring the most effective use of personnel and the application of the [United Nations] Common System conditions of employment." The Secretary General also undertakes administrative arrangements for the specialized secretariats of the permanent organs, and appoints staff to those secretaries in agreement with the head of each permanent organ. The secretariat of the permanent organs "shall work under the direct orders of those senior officials concerned but in accordance with general administrative directives of the Administrative Council and of the Secretary-General." The Secretary-General may also temporarily reassign staff members from each of the permanent organs to meet changing work requirements but must do so in consultation with the heads of the other permanent organs and must report all such reassignments to the Administrative Council. The Secretary-General must also report to the Administrative Council "any decisions taken by the United Nations and the specialized agencies which affect the [United Nations] Common System conditions of service, allowances and pensions."[3]

---

[3]Ibid., Art. 56 1., (d) & (g).

The Secretary-General's involvement with conferences and meetings is "to undertake secretarial work preparatory to, and following the conferences of the Union." The Secretary-General is also instructed to provide the secretariat for conference of the Union, in consultation with the inviting government, and for meeting of the permanent organs. He may also provide the secretariat of other telecommunication meetings on a contractual basis when requested."[4]

The Convention requires the Secretary-General to bring together, publish, and disseminate information as follows:

1. Official lists of data supplied by the permanent organs;
2. Principal reports of the permanent organs including recommendations and operating instructions;
3. Records of international and regional telecommunication agreements reported to him;
4. Technical standards and other data concerning assignment and utilizations of frequencies prepared by the IFRB;
5. A record of the composition and structure of the Union, the general statistics and official service documents of the Union, and other documents as conferences or the Administrative Council may direct;
6. National and international data concerning telecommunications throughout the world;
7. Technical and administrative information useful to developing countries in order to help them to improve their telecommunication networks;
8. Information "as would be of assistance to Members regarding the development of technical methods with a view to achieving the most efficient operation of telecommunication services and specially the best possible use of radio frequencies so as to diminish interference;" and
9. A journal of general information and documentation concerning telecommunication.

As if this was not enough to do, there is a catch-all phrase in the Convention which requires the Secretary-General to "perform all other secretarial functions of the Union."[5]

It should also be noted that the Secretary-General is the legal representative of the ITU and that he or his Deputy Secretary-General

[4]Ibid., Art. 56, 1., (i) and (j).
[5]Ibid., Art. 56, 1., (r) and (aa).

may participate in plenary assemblies of the International Consultative Committees and in all conferences and meetings of the Union in a consultative capacity.[6]

The Secretary-General and the Deputy Secretary-General

To many, the Secretary-General is the ITU — not the chairman of the plenipotentiary which has jurisdiction over the ITU's basic document, nor the chairman of the Administrative Council which is the legal agent of the plenipotentiary. The Secretary-General is always ready to greet visitors at the headquarters in Geneva. At plenipotentiaries and administrative conferences the familiar face at the podium is not the newly elected chairman, but the Secretary-General. He has already initiated the proceedings prior to the selection of the chairman and a number of documents with his name attached are the first things the delegates see on arriving at the conference site. Further, his name appears on numerous documents and messages that are in the hands of delegates long before the conference has even started. It is the Secretary-General who has the most influence in establishing the tone of the ITU. It is no wonder, therefore, that the Secretary-General is the physical symbol of the ITU.

The ITU has had six Secretaries-General since the post was created by the Atlantic City Plenipotentiary Conference of 1947:

1. Franz von Ernst (Switzerland), 1948-1949;
2. Leon Mulatier (France), 1950-1953;
3. Marco Aurelio Andrada (Argentina), 1954-1958;
4. Gerald C. Gross (U.S.A.), 1958-1965;
5. Manohar Balaji Sarwate (India), 1966;
6. Mohamed Mili (Tunisia), 1966 to present;

The Administrative Council, which until 1959 elected the Secretary-General, chose Franz von Ernst, who had been Director of the Bureau since 1935, as the first Secretary-General. Von Ernst served for two years when he retired and was replaced by Leon Mulatier who had been a Vice-Director of the Bureau. Leon Mulatier served from 1950 to 1953 at which time he was replaced by Andrada who

---

[6]Ibid., Art. 9, 3. and 56, 2.

had served on the Administrative Council and as Chairman of the Buenos Aires Plenipotentiary Conference of 1952. Andrada served in the office from 1952 until his death in 1958. In 1958 the Assistant Secretary-General, Gerald Gross, was chosen by the Administrative Council to serve until he was elected in his own right by the Geneva Plenipotentiary of 1959. On the announcement of the retirement of Gross, Manohar Sarwate, a Vice Chairman of the 1959 World Administrative Radio Conference, was elected Secretary-General and Mili the Deputy Secretary-General by the 1965 Montreux Plenipotentiary. Mohamed Mili was called upon by the Administrative Council to assume the duties of Secretary-General after the death of Sarwate, and Richard E. Butler (Australia) was elected by the 1968 Administrative Council to assume the post of Deputy Secretary-General to replace him. The 1973 Plenipotentiary Conference elected Mili Secretary-General and Butler Deputy Secretary-General. Both individuals are up for re-election at the 1982 Plenipotentiary.

Perhaps the most contested election was that of Gerald Gross in 1959. At the outset, there were six candidates in addition to Gross. On the first ballot the highest number of votes was garnered by Jean Rouviere (France), the Director of the CCITT. On the second ballot, the vote for Rouviere had risen to twenty-two, but that for Gross had risen to twenty-three. Between the second and third ballots, the candidates from Colombia, New Zealand, and Pakistan were withdrawn, leaving only Gross, Rouviere, and Shoukry Abaza, a citizen of the United Arab Republic. On the third ballot Gross was elected by fifty-one votes to thirty-five for Rouviere and none for Abaza.[7] In 1973 Mili had only one opponent and was elected on the first ballot by 104 votes; his rival received only twenty-three votes.[8]

---

[7]See ITU, *Documents of the Plenipotentiary Conference of the International Telecommunication Union, Geneva, 1959, Minutes of the Plenary Meetings.* Doc. No. 380-E, 11 December 1959, pp. 160-164 and ITU, *Minutes of the Plenipotentiary Conference of the International Telecommunication Union, Malaga-Torremolinos, 1973,* Geneva, 1974, Document No. 176-E, 14 October 1973, p. 230. At the next meeting, Richard Butler defeated Henryk Backo (Poland) in one ballot by 78 votes to 50.

[8]Before his election to Deputy Secretary-General Butler participated in a number of meetings of the Administrative Council and conferences of the Union as a member of the Australian delegation.

All of the Secretaries-General were telecommunications experts before their election to the office with the exception of Andrada and von Ernst. Mulatier and Gross had also served as Vice Directors of the Bureau and as Assistant Secretaries-General. Sarwate and Mili both served as Deputy Secretaries-General and Andrada, Sarwate and Mili all served on the Administrative Council and in ITU sponsored conferences and meetings before being elected Secretary-General.

The Secretary General is assisted by a Deputy Secretary-General. From 1947 to 1959 the ITU had two Assistant Secretaries-General but during that period it was found that one assistant tended to assume control of radio matters and the other telegraph and telephone matters, which tended to diminish the overall authority of the Secretary-General. Having two assistants of equal authority also raised problems concerning succession to the post of Secretary-General. Although there were proposals from a number of countries that would have eliminated any elected office subordinate to the Secretary-General, in view of the above reasons and for economy purposes the 1959 Plenipotentiary agreed that the number should be reduced to one and the name changed to Deputy Secretary-General.

The Deputy Secretary-General has three official duties: 1) to take over the post of Secretary-General if it falls vacant;[9] 2) to perform the duties of the Secretary-General in the absence of that individual; and 3) "to assist the Secretary-General in the performance of his duties and undertake such specific tasks as may be entrusted to him by the Secretary-General."[10]

In practice the specific function of the Deputy Secretary-General depend on his individual qualifications, the needs of the service, and especially the trust placed in him by the Secretary-General. The present Deputy has been particularly active in matters of ITU computer facilities, development assistance policy and defending the ITU's position *vis a vis* other international organizations.

---

[9] If the post of Deputy falls vacant more than 180 days before the date set for the convening of a Plenipotentiary, the Administrative Council appoints a successor for the balance of the term.

[10] ITU, *International Telecommunication Convention, Malaga-Torremolinos, 1973,* Art. 9,1., 3., and 4.

The Staff

The staff of the ITU's Secretariat is not large compared to many of the UN specialized agencies.[11] By January 1981, the total of the General Secretariat staff, including the two elected officials was 476. When added to the specialized secretariats of the IFRB and the CCIs the total came to 652.

There are four categories of staff: (1)Elected Officials; (2) Senior Counsellors (D); (3) Professional (P); and (4) General Service (GS). Each member of the regular staff has either a permanent contract or a fixed term contract.[12] The duties and responsibilities of the non-elected staff, as well as its salaries, are determined by the Secretary-General on the basis of the standards set by the United Nations Common System (and in conjunction with the Directors of the CCIR and the CCITT in regards to the specialized secretariats).[13] The duties of the elected officials are set by the convention and regulations. The convention provides that salaries of the elected officials be set at "an adequate level above those paid to appointed staff in the United Nations Common System." The levels established by the Malaga-Torremolinos Plenipotentiary are 124 percent for the Secretary-General, 111 percent for the Deputy Secretary-General and the directors of the consultative committees, and 106 percent for the members of the IFRB.[14] For the year 1981 the Administrative Council determined this to amount to a yearly salary of $55,614.16 for the Secretary-General, $49,783.63 for the Deputy Secretary-General and the Directors of the consultative committees, and $47,541.24 for the

---

[11] In 1976, for example, only four of the UN Specialized Agencies had fewer employees than did the ITU: the World Meteorological Organization, the Universal Postal Union, the Intergovernmental Maritime Consultative Organization, and the World Intellectual Property Organization. In that year the Food and Agricultural Organization alone had 3,303 permanent employees. See U.S., 94th Congress, 1st session, United States Senate, Committee on Government Operations, *U.S. Participation in International Organization*, U.S. Government Printing Office, Washington D.C., 1977, pp. 117-119.

[12] See Table 7/2 for a breakdown of the staff as of 1 January 1981.

[13] For additional information concerning employment practices and conditions in the ITU, see ITU, *Staff Regulations and Staff Rules of the International Telecommunication Union*, Geneva, 1969 as amended.

[14] See ITU, *International Telecommunication Convention, Malaga-Torremolinos, 1973*, Resolution Number 2.

members of the IFRB.[15] These salaries are, of course, tax free. The Secretary-General is also given a representation allowance of 15,000 Swiss francs, the Deputy Secretary-General and Directors of the CCIR and the CCITT each have a one of 7,500 Swiss francs, and the IFRB as a whole has one of 7,500 Swiss francs to be spent at the discretion of the chairman.

As with other secretariats, the professional staff of the ITU General Secretariat must be recruited with special reference to geographical distribution.[16] At the end of 1980 the headquarters staff included some forty-eight different nationalities; twenty-five individuals came from the Americas, 107 from Western Europe, ten from Eastern Europe and Northern Asia, thirteen from Africa, and twenty-nine from Asia and Australasia. The country with the largest single contingent, as might be expected, was Switzerland with thirty-six, followed by neighboring France with twenty, and the United Kingdom with sixteen. In the Americas, the United States was first with thirteen; in Eastern Europe, the U.S.S.R. had five; and in Asia, India had seven and Japan five. In Africa, Morocco and Tunisia were tied with three apiece.[17]

The regular headquarters staff is the foundation of the ITU General Secretariat, but is far from the total of the individuals who work in the ITU buildings in Geneva. As with many other international organizations, it is not possible to keep a permanent staff, especially those in the linguistic section, large enough to take care of the peaks of work that occur when conferences and meetings take place. Thus, almost every year the Secretary-General is forced to hire numerous part-time staff. In 1980 it was necessary to negotiate 1,183 short-

---

[15]These salaries are for individuals with dependents. If they have no dependents, the salaries are $50,476.98, $47,389.44, and $43,403.44 respectively. See ITU, Administrative Council, 36th session, Geneva, June 1981, Document 5682-E (CA36-101), 9 June 1981.

[16]The principle of geographical distribution applies to elected officials and staff in the professional category and above and in the technical posts in grades G.7, G.6, and G.5. The principal does not apply to grades G.4 and below.

[17]The figures in this section are from ITU, Administrative Council, 36th Session, Geneva, June 1981, Addendum No. 2 to Document No. 5621-E (31 March 1981). See also the article by Michael Bardoux, Chief of the ITU Personnel Department, "Geographical distribution and career service in the ITU," *Telecommunication Journal*, Vol. 45, No. V (May 1978), pp. 217-218.

term contracts which totalled some 62,621 working days because of the Regional MF Broadcasting conference (Region 2), the 35th session of the Administrative Council, and numerous CCITT and CCIR meetings. The largest number of contracts let out in 1980 were for interpreters (473). Others included staff necessary to carry out document production and distribution.[18]

Two interrelated problems have plagued the ITU Secretariat almost from the very beginning: how to attract individuals with the necessary qualifications and how, at the same time, to keep the geographical distribution high. In the early years of the new Secretariat, the task was that of turning a primarily Swiss staff into an international staff without at the same time making Swiss employees feel that they were not an important part of the system. This was accomplished over the years by increasing the overall size of the Secretariat and through attrition.

For many years, the ITU also subscribed to the concept that had surfaced in the League of Nations, that by offering recruits the possibility of becoming a part of the permanent staff, one could attract the best people and assure the impartiality of the Secretariat. In the early years the only objectors to the process were the Russians and their allies who argued in the ITU and elsewhere that it was impossible to make an individual "international" and thus members of the secretariat should consist mainly of individuals on short-term loan from member administrations. The Soviets were a small minority and it was only their nationals who came and went without becoming a real part of the permanent secretariat.

The influx of new, developing nations which reached its peak in the 1960s brought allies to the Russian position. These new nations wanted to be represented on the secretariat for prestige purposes, but in many cases did not have sufficient trained personnel to be able to send technical experts to the ITU. However, service on the ITU Secretariat was considered to be training for individuals which they could put to good use when they returned to their home administrations. For these reasons, many of the new, developing nations

---

[18]ITU, Administrative Council, 36th Session, Geneva, June 1981, Addendum No. 2 to Document No. 5621-E (31 March). Not included in either category being discussed is this section are the 299 experts whose salaries were paid for by the UNDP.

began to demand that a larger portion of the ITU Secretariat be on contracts rather than a part of the permanent staff. Although those who lean towards the permanent type of secretariat have prevailed,[19] the portion of members of the secretariat on fixed term appointments has begun to grow. In 1969, out of a total headquarters staff of 433, only twenty-eight were on fixed term contracts. In 1980, however, 126 of the total headquarters staff of 652 were on fixed term appointments.[20]

Organization

As illustrated in Table 7.1, the General Secretariat of the ITU is organized into six departments: Personnel, Finance, Conferences and Common Services, Computer, External Relations, and Technical Cooperation. There is also an Archives Service.

The Department of Conferences and Common Services is the largest and contains seven divisions: Language; Typing and Text Composition; Publications and Reprography; Dispatch/Transport Service; Conference Documents Service; Building, Supplies and Stores; and a Telephone Service. The work of this division also tends to increase in years of numerous conferences and meetings. In 1979, for instance, the Language Division translated 7,404 pages into English, 21,450 pages into French, and 24,871 into Spanish. Of this, 23 percent, or 12,401 pages were translated for WARC 79 alone. In the same year, the Typing Pool produced 102,239 pages, 46,345 of which were for the WARC 79 Conference.[21]

---

[19]See for instance the report of the debates on the matter in U.S. Department of State, Office of International Communications Policy, *Report of the United States Representative on the Administrative Council of the International Telecommunication Union, 32nd Session, Geneva, Switzerland, May 23 - June 10, 1977,* TD Serial No. 86, Washington, D.C., 1977 (mimeo), pp. 2-3. See also ITU, *International Telecommunication Convention, Malaga-Torremolinos, 1973, Resolution, No. 6.*

[20] See ITU, Administrative Council, 36th Session, Geneva, June 1981, Addendum No. 2 to Document No. 5621-E (31 March 1981). The Soviet delegate to the 1981 Administrative Council made a number of criticisms of the ITU recruitment policy, including what he considered was the Secretary-General's failure to increase adequately the ratio of fixed-term contracts to permanent contracts, the "inadequate" representation of East European nationals in the secretariat, and the "disproportionate" number of Americans on the CCIR staff." *Report of the U.S. Mission in Geneva to the Secretary of State on the 36th Session of the Administrative Council, Geneva, 1-19 June 1981,* Incoming Telegram No. 05650.

[21]*ITU, Report on the Activities of the International Telecommunication Union in 1979,* Geneva, 1980, p. 13.

The Computer Department provides support services for all of the permanent organs of the ITU, especially for the IFRB and its work with the Master Frequency Register and studies to avoid harmful interference. In 1979 the ITU computer was used for a total of 3,024 hours; 50.7 percent by the IFRB; 41.6 percent by the General Secretariat, 6.1 percent by the CCIR; and 1.6 percent of the time by the CCITT.[22]

The Department of External Relations is the contact point of the ITU with the outside world. Not only does it contain the Public Relations Division but also divisions concerning the relations of the ITU with the United Nations and other international organizations, relations with members of the Union, the division dealing with the preparation of Administrative Council sessions and conferences, and a division dealing with legal matters. Also in this department is the staff which prepares for the periodic international telecommuncation trade fair known as TELECOM, the last of which was held in conjunction with the World Administrative Radio Conference in Geneva in 1979.[23] Also included is the central ITU library and Documentation Center.

The second largest is the Technical Cooperation Department, one of those hybrid creations peculiar to international agencies affiliated with the United Nations Development Program. Although the ITU provides an invaluable service to member countries in the area of technical cooperation, it has little to say about the projects to be undertaken and the money that will be spent. As will be explained in some detail in Chapter Twelve, the decisions are made by the UNDP and the developing countries involved; the ITU's principal task is to execute those decisions. In this sense, the Technical Cooperation Department is simply a sub-contractor for the UNDP. For this the ITU receives a payment which reimburses the ITU for most of the expenses entailed by this department, including the salaries of the members of the General Secretariat who are assigned to it.

The Secretariat of the ITU is completed by an Archives, Communications and Microfilm Section. In addition to maintaining the ITU's extensive archives that date back to the 1860s, it also contains the mail and telex services for the entire ITU headquarters.

---

[22]Ibid., pp. 14-15 and 38.

[23]See below, Chap. 13.

The Coordination Committee and the Problem of Federalism

One additional organ of the ITU is provided for in the convention, although it is not designated as one of the permanent organs: that is the Coordination Committee, made up of the Deputy Secretary-General, the Director of the CCITT, the Director of the CCIR, and the Chairman of the IFRB, with the Secretary-General as its chairman. The Coordination Committee is the result of, and symbolizes, the problem of federalism that has plagued the ITU from its reorganization in 1947.

The ITU's federalism has both an historical and a functional rationale. The International Telephone Consultative Committee came into existence outside the Telegraph Union as a body which did research primarily on the basis of correspondence. While it was felt that it would be helpful to have such a body to undertake studies of special interest to the ITU, many felt that it should be as free as possible in the choice of other questions for study. Consequently, the CCIs were given considerable autonomy, in addition to the power to elect their own directors. Later, the IFRB, was given similar autonomy consonant with its proposed quasi-judicial role. Further, each of these bodies required a secretariat of at least a few specialists who would not necessarily be found in the General Secretariat. Almost from the beginning there were problems of conflicting authority.[24]

The potential for problems was recognized at the first meeting of the Administrative Council, during the debates on new ITU personnel regulations. The chairman of the council, Francis Colt de Wolf (U. S.

---

[24] At its 5th Session, for example, the Administrative Council was confronted with an irate Secretary-General who demanded that the Council affirm his right to make the final decision on hiring and firing staff except for the technical experts in the higher grades in the specialized secretariats of the CCIs and the IFRB. If the Secretary-General did not have the right to consider all other staff as a pool, a decision on the part of the Council to economize might involve "the necessity of dismissing temporary or supernumerary staff of the General Secretariat of very long standing of great value, while in the other organs, temporary of supernumerary officials would be kept, although perhaps less qualified and not so senior." Codding, *The International Telecommunication Union*, p. 434. As regards conflict of authority between the Secretary-General and the IFRB in particular, the best known case arose when the Administrative Council decided to acquire an ITU computer. The IFRB as the major potential user of the new facility chose a system which was contested by the Secretary-General, who pointed out that it was he who was responsible to the Administrative Council for the administration and finances of the Union, not the IFRB. See ITU, Administrative Council, 17th Session, Geneva, May-June, 1962, Doc. No. 2797/CA17-E (VI/1-1), 21 May 1962.

A.), stated that in order to assure a proper implementation of the personnel regulations, it would be necessary for the heads of the permanent organs to "meet frequently to discuss personnel questions in the same way as they discussed technical and administrative questions so as to ensure the necessary coordination and cooperation between the organs of the Union in the most effective manner." Georges Valense, the French delegate, pointed out that close cooperation was necessary to eliminate possible overlapping of jurisdictions on technical, administrative and budgetary questions and thus avoid "the division of the various organs of the Union into watertight compartments..."[25] Toward the end of the session the council adopted a resolution creating a Coordination Committee "of a purely advisory character" comprised of the Secretary-General, the Chairman of the IFRB, the directors of the CCIs and the vice-director of the CCIR "for the coordination of general questions involving the finances and the staff of the Union and other questions of common interest to the various permanent organs of the Union" which should "meet as frequently as it may be necessary."[26]

The 1959 Geneva Plenipotentiary gave the matter serious consideration, having before it examples of problems that had confronted the Administrative Council, and a report by the United Nations Advisory Committee on Administrative and Budgetary Questions in which it was stated: "while the Coordination Committee serves a limited but useful purpose, the basic problems which the complexities of the legislative and secretarial structure (of the ITU) entails do not seem to have been adequately met under the existing arrangements from the point of view of the sound and economic administration of the Unions's activities."[27] After lengthy discussions it was decided to retain the committee's basic structure but to reinforce its role as a coordinating body. This was done by inserting a paragraph in the Convention providing that it was the duty of the Secretary-General to use the Coordination Committee to "coordinate the activities of the permanent organs of the Union," and spelling out that

---

[25]See ITU, Administrative Council, Third Session, September, 1948, *Minutes of the 20th Plenary*, pp. 3 and 6.

[26]See ITU, Administrative Council, *Resolutions of the Administrative Council of the International Telecommunication Union (4th Session, Geneva, August-September, 1949)*, Geneva, 1949, *Resolution No. 48*, p. 22.

[27]See ITU, Plenipotentiary Conference of the International Telecommunication Union (Geneva), 1959), Doc. No. 8-E, 29 September 1959.

this coordination should encompass "administrative matters, technical assistance, external relations, public information and any other important matters laid down specifically by the Administrative Council."[28]

The 1965 Plenipotentiary was convened after problems had arisen over the proper organ to choose a computer system for the ITU and a number of proposals were submitted to it to change the status of the Coordination Committee. In the discussions that ensued, the basic issues raised by the federal structure of the ITU were aired. The delegate of the USSR had the most comprehensive criticisms and suggestions for change, and, although the Soviets may have had other motives connected with their attempt to reduce the authority of the Secretary of the United Nations, they did hit at the heart of the problem. According to the Soviet delegate, the Coordination Committee was not doing its job, thus making work for the Administrative Council. The committee should be given specific duties to perform such as relieving the Administrative Council of much of the work involved in technical cooperation and establishing the salaries of individuals in the general service category. Another delegate suggested that the committee be elevated to the status of permanent organ of the ITU and be given its own charter. The committee, it was argued, should have the power to decide on many questions of its own and not just make recommendations to the Administrative Council. The opponents of strengthening the committee came mainly from those who felt that, if it was given substantive power, the committee would infringe on the powers of the Secretary-General. A good organization, it was argued, could not be run by a committee.[29]

The result was a compromise. Although not made a permanent organ, the Coordination Committee was given special status by having its own article in the convention and its duties spelled out in more detail. The committee was designated to help the Secretary-General in the preparation of the *Telecommunication Journal*, the annual budget, the financial operating report and accounts and the annual report on the activities of the Union. The committee was

---

[28]See ITU, *International Telecommunication Convention, Geneva, 1959*, Art 10, 2., a.

[29]See ITU, The Plenipotentiary Conference of the International Telecommunication Union (Montreux, 1965), Doc. No. 305-E, 18 October, 1965, pp. 2-6; Doc. 419-E, 28 October 1965, p. 1; Doc. 445-E, 1 November 1965, pp. 1-5; and Doc. No. 493, 4 November 1965, pp. 4-5.

given independent powers to examine the work of the Union in technical cooperation, to submit recommendations to the Administrative Council and to coordinate ITU activities with other international organizations.

The 1965 Plenipotentiary made several decisions, some of which appeared contradictory in nature. On the one hand, the concept of federalism was given a boost by the 1965 Plenipotentiary when it decided that the committee should reach its conclusions by unanimous agreement. The Secretary-General's power was enhanced, on the other hand, by permitting him to ignore the unanimity principle and make decisions himself when "he judges the matters in question to be of an urgent nature."[30] When this is the case, and the committee insists, he reports the matter to the Administrative Council "in terms approved by all the members of the Committee."[31] If the matters are deemed to be important but not urgent, they are to be referred to the next session of the council. In the event that the Secretary-General might refuse to call meetings of the committee it was also decided that it should, "in general," meet at least once a month.

The 1973 Malaga-Torremolinos Plenipotentiary Conference made further modifications to the article dealing with the Coordination Committee which tended to uphold the federal principle and limit the autonomy of the Secretary-General. The 1973 Plenipotentiary confirmed the right of the Secretary-General to make urgent decisions but specified that if he did so he must report promptly in writing on such matters to the members of the Administrative Council, setting forth his reasons for such actions together with any other written views submitted by other members of the committee." The 1973 Plenipotentiary also felt that holding meetings once a month "in general" was ambiguous and could lead to abuse. It attempted to reinforce the principle of monthly meetings by stipulating that the committee should meet "normally" once a month.[32]

---

[30]ITU, *International Telecommunication Convention, Montreux, 1965*, Geneva, 1965, Art. 11, 2. The power of the Secretary General to make decisions on his own is conferred by the interesting sentence: "The Secretary General may . . . take decisions even when he does not have the support of two or more other members of the Committee . . . "

[31]Ibid.

[32]See ITU, *International Telecommunication Convention, Malaga-Torremolinos, 1973*, Geneva, 1974, Art. 59.

In actual practice, the Coordination Committee does meet once a month on schedule for a short meeting. From the point of view of some the committee seldom if ever discusses anything of substance at these meetings but simply acts as a "rubber stamp" for decisions that the Secretary-General has already made. The Coordination Committee, in other words, is not doing its job, according to those individuals; rather it is being used as an instrument to strengthen the hand of the Secretary-General over the heads of the other permanent organs. The Secretary-General, however, is of the opinion that there are fewer and fewer issues on which the members of the committee have diverging views, therefore it is only normal that the meetings are short and perfunctory.[33]

It would be almost impossible to tell which of these views is the correct one. Nevertheless, as long as the ITU subscribes to the concept that there should be a number of semi-autonomous permanent organs, there will be a need for a body such as the Coordination Committee if the Administrative Council is not to take on all of the job of continually sorting out all the jurisdictions of the permanent organs of the ITU and making decisions on personnel matters which under other international organs would be taken care of by the Secretary-General in the normal course of his duties. The problem is compounded by the fact that the directors of the CCIs are elected by their own plenary assemblies which are quite different constituencies than that reflected in the normal plenipotentiary conference.

Conclusions

The work of the Secretariat of the ITU is much less glamorous than that of the delegates from national administrations to treaty-making conferences and meetings. It is, however, essential to the ITU's general mission. Indeed, without its Secretariat, the ITU would cease to exist.

---

[33]From the point of view of the Administrative Council, too many problems are still coming to it that should be solved in some other fashion. In this respect, it should be noted that the computer issue surfaced again at the 1981 Session due to a decision of the 1979 Administrative Radio Conference to computerize additional IFRB activities. The discussion led the U.S. Delegate to report that "doubts were expressed . . . as to the capability fo the IFRB and General Secretariat to implement the decision, particularly in view of history of difficult relations between (the) two staffs . . " *Report of the U.S. Mission in Geneva to the Secretary of State on the 36th Session of the Administrative Council,* Incoming Telegram, 05650.

The task of the ITU Secretary-General is not an easy one. There are a number of additional troublesome problems that need continuous attention including that of maintaining a highly qualified staff, providing for a broad geographical distribution in the higher grades, coping with the conflicting concepts of career employment versus short-term contracts, and overcoming the difficulties caused by fluctuations in national currencies. All of these problems are shared by the many other international organizations which have their headquarters in Geneva and the solutions to them must be approached on a broad basis.

The federalism that exists among the ITU's organization gives rise to special problems. It was believed that there were merits in giving the consultative committees, and later the IFRB, a degree of autonomy in structure and operation. As long as the members of the ITU continue to feel that this autonomy has some benefits, the problem of federalism will remain a problem that the various organs of the ITU, especially the Administrative Council, must be prepared to live with on a continuing basis, the argument of the Secretary-General that consolidation of all of the secretariats under his complete authority would be the most efficient solution notwithstanding. The Coordination Committee approach seemed to have a great deal of potential on paper, but does not seem to have worked in practice. Perhaps the functions of the committee could be redefined in such a way as to make it a viable entity or another type of coordinating mechanism could be subsituted for it.

TABLE 7.1

**THE GENERAL SECRETARIAT OF THE
INTERNATIONAL TELECOMMUNICATION UNION**

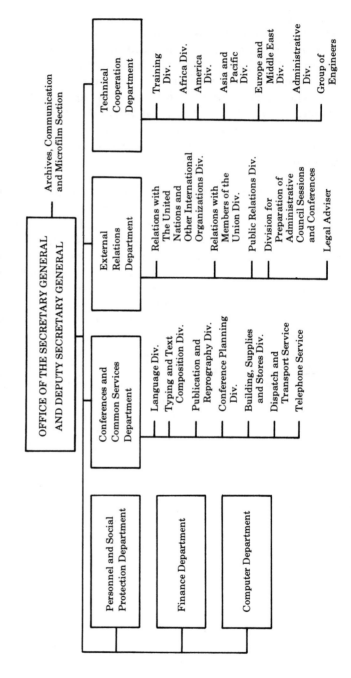

TABLE 7.2
DISTRIBUTION OF STAFF ACCORDING TO GRADE

Grade	General Secretariat		IFRB		CCITT		CCIR		Total	
	Perm.	FT	Perm.	FT	Perm.	FT	Perm.	FT	Perm.	FT
Elected officials	2		5		1		1		9	
D.1	7	2	2	–	3	–	3	–	15	2
P.5	15	4	6	–	3	1	5	–	29	5
P.4	30	25	14	4	11	–	6	–	61	29
P.3	35	5	4	–	2	–	–	–	41	5
P.2	6	4	9	–	1	–	–	–	16	4
P.1	–	–	–	–	–	–	–	–	–	–
G.7	46	2	13	–	5	–	5	–	69	2
G.6	51	11	12	–	15	1	8	–	86	12
G.5	68	42	13	–	1	–	2	–	84	42
G.4	34	11	12	–	1	–	1	–	48	11
G.3	49	11	5	–	–	–	–	–	54	11
G.2	10	3	1	–	–	–	–	–	11	3
G.1	3	–	–	–	–	–	–	–	3	–
Total	354	120	91	4	42	2	30	–	517	126
	476		100		45		31		652	

(Perm. = Permanent; FT = Fixed-term)

Note: The above table shows the grades of staff members, those receiving a special post allowance granted upon a recommendation by the Appointment and Promotion Board being shown in the grade for which they receive the allowance.

chapter 8

FINANCES

As the scope of the work of the ITU has broadened over the years and as the number of members has increased, the amount of money that is spent by the ITU has also tended to increase. Still the expenses of maintaining the ITU do not appear to be out of line with those of comparable international organizations.

Expenses

The expenses of the ITU fall into two major categories: those involved in the work of the Administrative Council, the General Secretariat, the IFRB and the CCIs, and those resulting from the holding of plenipotentiary and administrative conferences. The upper limits of both are established by the plenipotentiary conference and are valid until the next plenipotentiary.[1] The ITU also has a separate budget for development assistance, which is met by UNDP, some other outside funds, and a publications budget which is supposed to be self-financing. All three budgets are included in a consolidated report.

The 1973 Malaga-Torremolinos Plenipotentiary Conference set the limits to the expenses for the Administrative Council and the permanent organs for 1974 to 1979 in Swiss francs as follows:

1974 — 35,000,000	1977 — 37,600,000
1975 — 36,650,000	1978 — 38,800,000
1976 — 36,600,000	1979 — 39,980,000

---

[1] See ITU, *International Telecommunication Convention, Malaga-Torremolinos, 1973*, Arts 6, 2, c.

In anticipation that the next plenipotentiary would not be held in 1979 as provided in the convention, the 1973 Plenipotentiary also decided that budgets after 1979 could be raised by a maximum of 3 per cent per year.

The 1973 Plenipotentiary established the following maximum limits (in Swiss francs) for the expenses of conferences and meetings of the CCIs:[2]

1974 — 6,600,000	1977 — 3,400,000
1975 — 2,900,000	1978 — 3,000,000
1976 — 11,000,000	1979 — 14,800,000

The limits set by the plenipotentiary conferences are not absolutely rigid, however. The Administrative Council is authorized to exceed the limits to take account of increases in the salary scales and pension contributions or allowances, including post adjustments established in the United Nations Common System, and of fluctuations in the exchange rate between the Swiss franc and other currencies which involve additional expenses to the ITU. The Administrative Council is also authorized to exceed the limits set for conferences and meetings if the excess involved can be compensated by sums which are accrued from a previous year or are foreseen for a future year. If all else fails, and the Administrative Council is convinced that the limits to both types of expenditures are "insufficient to ensure the efficient operation of the Union," it may exceed those limits with, and only with, the approval of a majority of the members of the ITU and after the Council has presented a full statement of the facts justifying that step.[3]

With this in mind, we can proceed to an overview of the expenses of the ITU since 1960, when a consolidated budget was introduced, as shown in Table 8.1.

The overall increases in expenditures beginning in 1966 and 1974 reflect the raising of the limits by the 1965 and 1973 Plenipotentiaries respectively. The sporadic increases prior to 1974 were due to the

---

[2] See Ibid., *Additional Protocol I,* Expenses of the Union for the Period 1974 to 1979. It was also provided that if a Plenipotentiary Conference was not held in 1979 as anticipated, the expenses for conferences in 1979 would be reduced by 3,800,000 Swiss francs; this subsequently occurred.

[3] Ibid.

holding of major conferences, e.g., in 1963, 1965, 1966, 1970, 1971 and 1973. Since 1974 almost every year has seen the holding of an important ITU conference.

The summary of the accounts for the Union in 1979 will give us a good idea of how the money is allocated:[4]

	Amount in Swiss Francs	%
1. Common headquarters expenditure ...	49,720,760	78.3
2. Conferences (WARC-79)	5,318,917	8.4
3. CCITT and CCIR Meetings	2,645,056	4.2
4. Administrative Council Meeting	708,932	1.1
5. Common expenditures for conferences and meetings	4,410.775	6.9
6. Other (including 131,276 for "writing off bad debts")	716,870	1.1
TOTAL	63,521,310	100

As might be expected, the largest single item under "Common headquarters expenditure," and the largest single item in the entire budget, was for personnel. In 1979, it amounted to 43,331,674 Swiss francs, 35,532,178 for salaries and 7,799,497 for "social security." Buildings and office expenses amounted to another 5,742,787 Swiss francs. In the item "CCITT and CCIR Meetings," only 120,164 was attributed in 1979 to the CCIR.

Income

As is the case with most international organizations, the ITU receives the major portion of its income from contributions by member countries. Unlike other international organizations, the ITU also receives a contribution from the private operating agencies that it allows to participate in conferences and meetings of the consultative committees.

The major source of income is, of course, the contributions that are received from the member countries. The amount that each member

---

[4]See ITU, *Report of the Activities of the International Telecommunication Union in 1979,* Geneva, 1980, p. 130.

country contributes to the ITU's budget depends upon the contributory class that it selects according to the following scale:

30 units	13 units	3 units
25 units	10 units	2 units
20 units	8 units	1½ units
18 units	5 units	1 unit
15 units	4 units	½ unit

Members are free to choose any class of contribution, although a resolution of the Montreux Plenipotentiary suggested that it would be helpful if members choose a class of contribution "most in keeping with their economic resources."[5]

Of the 154 members of the ITU on December 31, 1980, thirty-eight were in contributory classes 1½ to 30 and all of the remainder in classes 1 and ½ (eighty-two in the ½ class). The countries in classes 5 to 30 were as follows:[6]

Class 30: France, U.K., U.S.A., and U.S.S.R.
Class 25: Germany (F.R.)
Class 20: China and Japan
Class 18: Australia and Canada
Class 15: none
Class 13: India
Class 10: Italy, Netherlands, Sweden, and Switzerland
Class 8: South Africa
Class 5: Belgium, Brazil, Denmark, and Norway

Classes 1 and ½ were dominated primarily by the developing countries. However, they also contain most of the members of OPEC.

All of the expenses of the ITU which are not covered from other sources are divided by the total number of contributory units to

---

[5]ITU, *International Telecommunication Convention Montreux, 1965*, Geneva, 1965, Resolution No. 15.
[6]See ITU, Administrative Council, 36th Session, Geneva, June, 1981, Annex 1 to Doc. No. 5621-E (31 March 1981).

obtain the amount of a single contributory unit. In 1980 a total of 426½ units were subscribed to by member countries. After applying the 426½ subscribed units to the budget estimates for 1979 of 53,909,600 Swiss francs, the Administrative Council at its 1979 meeting fixed the amount of one contributory share for 1979 at 126,400 Swiss francs. At this rate, those member countries which had chosen the maximum class were obliged to pay a total of 3,792,000 Swiss francs while those in the lowest class paid only 63,200 Swiss francs towards the expenses of the Union in 1979. In an interesting reversal from the situation that exists in most international organizations, in the ITU the largest single contributor is the U.S.S.R. While the U.S.S.R. has chosen a 30 contributory class for itself, as has the United States, it also subscribes to a class 1 contribution for Byelorussia and a class 3 contribution for the Ukrainian S.S.R.

The second source of income for the International Telecommunication Union, albeit a small one, is contributions from private operating agencies, scientific or industrial organizations, and other international organizations. As was discussed earlier, private operating agencies which have the approval of a member government may participate as members of the CCIR and the CCITT. Regional international telecommunication organizations and scientific or industrial telecommunications organizations may participate in an advisory capacity. In addition, the United Nations, its Specialized Agencies, the International Atomic Energy Agency, and regional international telecommunication organizations may send observers to plenipotentiary conferences; and those organizations plus other international organizations and recognized private operating agencies may participate as observers to administrative conferences. In principle, all must share in the expenses of those conferences and meetings in which they participate and each must choose a class of contribution from the regular contribution scale. For the expenses of consultative committees, the Administrative Council annually fixes the amount of the contribution unit. In 1980 it amounted to 21,100 Swiss francs. For other conferences and meetings, the amount the organization pays is based on the overall cost of that meeting.

Those entities classified as "international organizations" in the Convention, however, may be exempt from making a contribution.[7]

In 1979, forty-eight private companies subscribed to 45½ units of the CCIR's expenses and forty-eight companies subscribed to 49 units of the CCITT's expenses. American Telephone and Telegraph Co. and the United Kingdom Post Office shared the honors of being the largest single contributors to both CCIs with a 5 unit contribution. Canadian Telecommunications Carriers Association came next with a 4 unit contribution to both. Most companies, it must be noted, chose the ½ unit contribution. In the same year thirty-eight "Scientific or Industrial Organizations" contributed 25½ units to the CCIR's expenses and 135 contributed 79½ units to the CCITT's expenses. The largest single contributor was Siemens A.G., Munich, which contributed three units to each of the CCIs; Allgemeine Elektricitats-Gesellschaft, AEG-Telefunken Backnang, which contributed two units to each; and N. V. Philips' Telecommunicatie Industrie, Hilversum, which contributed one unit to the CCIR and

---

[7] In 1979, for example, there were twenty-five international organizations listed as exempt from contributing to the CCITT's expenses and one, the Intergovernmental Bureau for Informatics (IBI), as having presented a request for an exempt status. Twenty-two were exempt from contributing to the CCIR's expenses. The following were exempt from both: European Space Agency (ESA); International Air Transport Association (IATA); International Maritime Radio Association (CIRM); International Electrotechnical Commission (IEC); International Conference on Large High Tension Electric Systems (CIERE); International Federation for Documentation (FID); Ibero-American Television Organization (OTI); International Radio and Television Organization (OTI); International Radio and Television Organization (OIRT); African Postal and Telecommunications Union (APTU); Arab Telecommunications Union (ATU); Union of National Radio and Television Organizations of Africa (URTNA); European Broadcasting Union (EBU); Pan-African Telecommunication Union (PATU); and International Union of Radio Sciences (URSI). The following eleven organizations were registered as participating in but exempt from contributing to the CCIR's expenses: Inter-American Association of Broadcasters (IAAB); International Time Bureau (ITB); Committee on Space Research (COSPAR); International Special Committee on Radio Interference (CISPR); Inter-Union Commission on Allocation of Frequencies for Radio Astronomy and Space Science (IUCAF); International Council of Scientific Unions (ICSU); International Astronautical Federation (IAF); International Astronomical Union (IAU); Asia-Pacific Broadcasting Union (IARU). The following eight organizations were listed as participating in CCITT activities but exempt from contributing to its expenses: International Teletraffic Congress (ITC); International Federation for Information Processing (IFIP); International Organization for Standardization (ISO); International Criminal Police Organization (ICPO); International Gas Union (IGU); International Union of Railways (UIC); International Union of Producers and Distributors of Electrical Energy (UNIPEDE); and International Union of Public Transport (UITP). See ITU, Administrative Council, 35th Session, Geneva, May, 1980, Doc. No. 5454-E (CA35-22), 28 February 1980, pp. 132-136.

two units to the CCITT. Almost all the others chose the ½ unit category. Finally, four international organizations contributed ½ units to the CCIR and seven contributed ½ units to the expenses of the CCITT.[8]

In the 1980 accounts, this source amounted to an additional 1,523,596 Swiss francs for the CCIR and 2,857,404 for the CCITT or about 7 per cent of its total income.

Contributions are payable at the beginning of the following year. The ITU does not have a working capital fund but it does have an agreement with the Swiss government by which the Swiss stand ready to place any necessary funds in the hands of the Secretary-General if there is a shortfall. In recent years, most of the contributions of member countries have been paid early enough to cover the ITU's expenses without the need to call on the Swiss government. Some of the unused funds have actually been available for short-term investment.

As mentioned earlier, the ITU also has a separate budget for technical cooperation and for publications. The former is made up of the administrative costs involved in administering the ITU's portion of the annual UNDP technical assistance program and is reimbursed by the UNDP. For 1980 this was calculated to be 9,872,713 Swiss francs. As regards the publications budget, for 1980 the cost of publications was 6,642,043. Income was 6,683,489, leaving a profit of some 41,446 Swiss francs.

The accounts of the ITU are audited on a regular basis by the Swiss government. There is no charge for this service.[9]

---

[8]Those contributing to the CCIR were: European Space Agency (ESA); International Association of Lighthouse Authorities (IALA); Arab Satellite Communications Organization (ARABSAT); and International Telecommunications Satellite Organization (INTELSAT). Those contributing to the work of the CCITT were: European Computer Manufacturers Association (ECMA); European Computing Service Association (ECSA); International Telecommunications Users Group (INTUG); International Chamber of Shipping (ICS); International Press Telecommunications Council (IPTC); Arab Satellite Communications Organization (ARABSAT); and International Telecommunications Satellite Organization (INTELSAT). Ibid.

[9]For additional information concerning the ITU's budgetary practices, see ITU, *Financial Regulations, Edition 1979*, Geneva, 1979.

Arrears

As could be expected, not all of the members of the ITU pay their debts on time; thus, the organization has the continuing problem of accounts in arrears. Before 1973 the ITU Convention did not provide any sanctions against a member country that failed to pay its contributions on time other than an interest charge on the amount owed. According to the ITU, the methods employed to encourage countries to pay included:[10]

1. Sending reminder letters and telegrams;
2. Publishing a list of sums in arrears every quarter in the ITU Notifications;
3. Direct approaches to the diplomatic missions in Switzerland;
4. Direct approaches to the responsible officials during official missions in the countries in question;
5. Review of payments in arrears by the Administrative Council and the Plenipotentiary Conferences and appeals to members to settle them.

These measures did not prevent arrears, however. While the total arrears for the year 1960 was only 2,243,361.80 Swiss francs, it began to grow fairly rapidly in the late 1960s with the increases in budgets due to the increasing number of radio conferences. By 1970 it amounted to 8,930,963.17 Swiss francs and in 1972 to 11,168,411.71 Swiss francs.[11]

The 1973 Malaga-Torremolinos Plenipotentiary investigated the matter and decided to introduce a sanction. According to the 1973 Convention, any member in arrears in its payments to the Union will lose its right to vote in conferences, meetings, and consultations carried out by correspondence so long as the amount in arrears equals or exceeds the amount of the contributions due from it for the preceding two years.[12]

[10]From an undated, mimeographed ITU document entitled "Replies to the Questionnaire of the Ad Hoc Committee of Experts to Examine the Finances of the United Nations and the Specialized Agencies," Geneva, circa 1965, p. 27.
[11]See ITU, *Report on the Activities of the International Telecommunication Union, 1960*, p. 62; *1970*, p. 175; and *1972*, p. 164.
[12]ITU, *International Telecommunication Convention, Malaga-Torremolinos, 1973*, Art. 15, 7.

The number of debtor countries fell drastically in 1974, and the debt fell to less than one million Swiss francs, but since that time it has begun a new climb. In 1980, for example, arrears in contributions amounted to over five and a half million Swiss francs. The largest single debtor in 1980 was Sudan with an outstanding debt of 579,635.85 Swiss francs followed by Chad with a debt of 553,897.60 Swiss francs. Non-member debtors, including private operating agencies, owed an additional 663,787.30 Swiss francs to the total for 1980.[13]

The plenipotentiaries often reduce or even eliminate accounts in arrears if the circumstances so dictate. At the Buenos-Aires 1952 Plenipotentiary, for instance, the debts which were not paid during World War II for Yugoslavia, former Italian colonies, and former Japanese dependencies and territories under Japanese mandate were all written off. At the 1965 Plenipotentiary it was decided that the debt owed by San Marino, which had ceased to be a member of the Union on December 31, 1948, was too small (approximately 23,000 Swiss francs) to merit the time and effort of collection, and was also written off. The 1973 Plenipotentiary wrote off Nicaragua's contribution for 1973 and allowed it to reduce its 1974 contributory class to ½ unit because of a serious earthquake in December, 1972.[14]

The ITU's actions concerning accounts in arrears are not always completely logical, as demonstrated by its actions towards South Africa. The 1973 Plenipotentiary Conference, it will be recalled, stripped South Africa of its rights to participate in conferences and meetings. After this decision, South Africa, understandably, started witholding its annual contribution. Despite the fact that the only sanction that could be taken against a member for arrears was depriving the country of the right to vote in conferences and meetings, the ITU began dunning South Africa to make its con-

---

[13]See ITU, Administrative Council, 36th Session, Geneva, June, 1981, Addendum No. 1 to Document No. 5621-E, 27 April 1981, p. An. 6/2.
[14]See ITU, *International Telecommunication Convention, Buenos Aires, 1952, Resolution No. 12,* and *Montreux, 1965, Resolution No. 14,* and *Malaga-Torremolinos, 1973, Resolution No. 15.* Because they were having a difficult time in paying their accounts, the 1973 Plenipotentiary also made special payment arrangements for Bolivia, Chile, Costa Rica, Dominican Republic, El Salvador, Haiti, Peru, Uruguay, and the Yemen Arab Republic.

tributions. The farce ended in 1977 when South Africa cleared up its arrears and again started making its 8-unit annual contribution.[15]

Criticisms of the Contributory System

The ITU's class-unit contributory scheme has been modified a number of times over the years, mostly to extend the range of the classes downward in order to allow the developing country members to pay less. Recently there has been a move to change to the UN system in which the United States pays the largest amount.

The ITU started out in 1868 by establishing six classes of contributions to the expenses of the International Bureau: 25, 20, 15, 10, 5 or 3 units. By 1947 the membership of the ITU included some smaller, less developed countries and to take this into account, the 1947 Plenipotentiary added a new class of 30 units and one of 1 unit. To accommodate the large number of small and less developed countries that had joined the Union since 1947, the 1952 Plenipotentiary added a new ½ unit, and to allow the more developed countries more flexibility in finding a class-unit more closely adapted to their ability to pay, added 18, 13, 8, 4 and 2 unit classes. Finally, the 1973 Plenipotentiary added a 1½ unit class to give the less developed nations a similar flexibility but turned down the addition of a ¼ unit.

During the centennial plenipotentiary in Montreux in 1965, for the first time some open criticisms were heard over the class-unit system as a whole. Both Mexico and Belgium presented proposals, the general tenor of which was that it might be more equitable if the amount of the contributions better reflected the amount of telecommunications in each member state. The Mexican proposal would have raised the American contribution to almost fifty percent, according to the U.S. delegate.[16]

Because of the strong opposition of the United States, the conference did not pursue the subject further but did request that the Secretary-General study possible improvements in the financing of the Union

---

[15]See Department of State, Office of International Telecommunications Policy, *Report of the United States Representative on the Administrative Council of the International Telecommunication Union, 32nd Session, Geneva, Switzerland, May 23 - June 10, 1977,* TD Serial No. 86, Washington, D.C., p. 4.

[16]See U.S., Department of State, Office of Telecommunications, *Report of the United States Delegation to the Plenipotentiary Conference of the International Telecommunication Union, Montreux, Switzerland, September 14 to November 12, 1965,* TD Serial, No. 973, Document No. 69, Washington, D.C., p. 34.

and that the Administrative Council submit to the next pleni-potentiary conference specific suggestions for such improvements.[17]

The 1973 Plenipotentiary also saw for the first time a strong movement to completely change the system. The issue was raised by Yemen and Mexico who suggested that the ITU adopt the UN system, which is based on a country's GNP and the exact amount of which is set by a committee of the General Assembly.[18] Under this system, the United States is by far the largest contributor, making up 25% of the total of the annual UN budget. The proposal had support in committee from some other Latin American countries and a number of African and Arab members but was defeated in a vote by 42 to 7 with 5 abstentions.[19] The issue was raised again in a plenary meeting and an extensive debate was held on the matter. The conference had before it a report from the Administrative Council which discussed various alternative financing systems but did not propose any change from the existing system. The conference also had before it a proposal from some twenty developing nations that the ITU adopt the UN system in which the amount of contributions to ITU expenses would be fixed by a committee on the basis of the income of each member country. A number of developing countries argued in favor of the UN plan because they felt it was much more equitable than the ITU system and because, according to the Mexican delegate, the ITU system does not "take into account the acute situation of countries with weak economies which were liable to be submerged by debts due to excessively high contributory unit charges."[20] Besides, the developed countries gained more from the work of the ITU than many of the poorer countries of the world.

[17]See ITU, *International Telecommunication Convention, Montreux, 1965, Resolution No. 11.*

[18] In 1968, as a part of its ongoing attempt to standardize the financial regulations of the various units of the UN family of international organizations, the UN General Assembly had passed a Resolution (No. 2474 (XXIII)) which recommended: "that the specialized agencies . . .whose scales of contributions still reflect significant varia-tions from the United Nations scale should intensify their efforts with a v: w to bringing their scales into harmony with the United Nations scale at the earliest possible time, taking into account differences in membership and other pertinent factors." Only the ITU and the Universal Postal Union fell into this category.

[19]See U.S., Department of State, Office of Telecommunications, *Report of the United States Delegation to the Plenipotentiary Conference of the International Tele-communication Union, Malaga-Torremolinos, Spain, September 14 - October 25, 1973,* TD Serial No. 43, Washington DC, 1973 (Mimeo.) pp. 26-27.

[20]See ITU, *Minutes of the Plenipotentiary Conference of the International Tele-communication Union, Malaga-Torremolinos, 1973,* p. 470.

In addition to Mexico, the UN system was supported by delegates from Afghanistan, Argentina, Iraq, Peru, Somalia, and Zaire.

The proposal, not surprisingly, was opposed by the United States. The United States delegate simply declared that his delegation and the U.S. government were both unalterably opposed to any change and, if it was made in the form that was being advocated, the United States would make a reservation to it and there was the further likelihood that the U.S. government might refuse to ratify the new convention. The United States was supported in its desire to maintain the status quo by Australia, the Federal Republic of Germany, Italy, Japan, the Netherlands, and Switzerland. The Swiss argued that the actual ½ unit class contribution was not a great burden on any country, amounting as it did to some 35,000 Swiss francs, or about one-thousandth of the total expenditure of the Union. It would buy not much more than a travel and subsistance allowance for one or two delegates to the conference.

The United States was, surprisingly, supported strongly by the U.S.S.R., Byelorussia, and Poland, all of whom echoed the U.S. threat that any attempt by the Malaga-Torremolinos conference to adopt the UN system might well result in a refusal by their governments to ratify the new convention.

After a few legal maneuvers, the proposal to adopt the UN system was defeated by a vote of forty-eight in favor, sixty-seven opposed, and three abstentions.[21]

The plenary did, however, adopt another proposal which instructed the Administrative Council to continue to study the matter and to find a solution to the problem, taking into account the discussions at Malaga-Torremolinos, including:

(a) Widening of the range of classes of contributions chosen by each member while maintaining freedom of choice;
(b) Application of a system for calculating contributions based on regularly updated official data; for example, the United Nations scale, a percentage based on such factors as the international telephone traffic of each member country, number of telephones, and gross national product.[22]

---

[21]Ibid., p. 472.
[22]ITU, *International Telecommunication Convention, Malaga-Torremolinos, 1973, Resolution No. 8.*

The Administrative Council was directed to submit the results of this study to all members at least one year before the next plenipotentiary is held.

The ITU's Administrative Council discussed the matter at its 31st, 34th, and 35th sessions and came to the conclusion that there were only two possible choices: to retain the ITU class-unit scheme or to adopt the UN system where contributions were based on the members' GNP. With a few exceptions, the majority of the members of the Council favored the first of these two choices.[23]

The 36th Session in 1981 had before it a lengthy report from the Secretary-General on the matter summarizing the discussions on the pros and cons of the two methods and concluding that the ITU's system of free choice should be maintained as well as the 1 to 60 ratio (½ unit to 30 units) between the largest and smallest contribution.[24] The UN system, according to the report, had three major advantages:

1) Under the United Nations system, the level of contributions is kept constantly up to date by a Committee appointed specifically for the purpose. Under the ITU system, downward adjustments may be made only at a Plenipotentiary Conference.

2) The United Nations system provides for a much wider range of contributions, which means that the distribution of costs would be more in keeping with each country's economic and financial means.

3) No prestige attaches to the level of contributions, which is intended to be based on each individual Member's capacity to pay.[25]

The ITU system, however, had two major advantages which "were far from being matched by those of the United Nations system."

[23]A third possibility, that of determining the amount of the contribution based on the level of use of telecommunication in a country, was abandoned because of the difficulty of obtaining the data necessary to make such an evaluation, among other problems.

[24]The UN system would widen this ratio to 1 to 2,500 (the lowest contribution being 0.01% and the highest 25% in the UN system.) The contribution of the United States would increase from 7.01% to 25%. The U.S.S.R contribution would increase from 7.49% to 22.74%. The vast majority of members would drop from 0.12% to 0.01%.

[25]See ITU, Administrative Council, 36th Session, Geneva, June 1981, Document No. 5584-E (CA 36-3), 5 November 1980, p. 6.

Those advantages being:

1) Under the system applied by the ITU, the contributions of a few Members cannot be so large that they predominate over those of the other Members taken together. If that were so, a suspended or delayed payment by one of the main contributors would suffice to constitute a serious danger to the Union's financial situation. The ITU system also makes for greater equality among Members.

2) The provisions of the [ITU] Convention and the freedom to choose a number of contributory units enable RPOAs [Recognized Private Operating Agencies] and SIOs [Scientific or Industrial Organizations] to share in defraying the expenses of the Union, particularly those of the CCIs. This would not be possible under a system of fixed contributions, which would require RPOAs and SIOs to attend conferences and meetings as part of their national delegations. Furthermore, financial contributions by such bodies would have to be made as part of the national contributions.[26]

The report occasioned a discussion which revealed that the UN system had gained a great deal more support than it had at the 1980 meeting. In 1980 only France and Ethiopia had been vocal in their support of the UN system.[27] In 1981 France and Ethiopia were joined by Cameroon, China, Lebanon, Mexico, Romania, Tanzania, and Venezuela. The U.S. delegation was of the opinion that the increase in supporters of the UN system was due to the increase in expenses that the Council felt itself forced to authorize for 1981, especially that for the new computer capacity for the IFRB.[28] Whatever the reason, the Council decided that it would not be possible to enter into an extended debate on the relative merits of the two approaches at that time. As a compromise it deleted the conclusions to the report and requested the Secretary-General to immediately forward the remainder to the member administrations as required by the 1973 Plenipotentiary resolution.

---

[26]Ibid, pp. 6-7. The Report also contained the note: "It should be stressed that the RPOAs and SIOs make a useful contribution to the work of the CCIs and contribute their scientific and technical knowledge to conclusions and Recommendations with consequences on the telecommunications development."
[27]The French contribution to the ITU expenses would drop from 7.01% to 6.16% under the UN system. Ethiopia would drop from 0.23% to 0.01%.
[28]See Report of the U.S. Mission in Geneva to the Secretary of State on the 36th Session of the Administration Council (1981), Incoming Telegram No. 06268.

In so doing, the Administrative Council added one more volatile issue to the agenda of the 1982 Plenipotentiary.

Conclusions

The ITU has a fairly straightforward financial arrangement that seems to have served it well. As a result of the annual interaction between the Secretary-General and the Administrative Council, there have not been any special surprises over the years. The ITU is especially blessed in this regard by the willingness of the Swiss government to advance operating funds in case of necessity and to provide a free audit of the ITU's accounts.

The continual rise in the overall costs of the ITU does not seem excessive in view of the rather complicated structure of the ITU and the myriad tasks that it performs for its membership. Even if the ITU does not make any drastic changes in either its structure or functions, as long as the cost of services and materials continues to increase, the expenses of the ITU can also be expected to continue to rise.

Since the ITU, as is the case with most international organizations, does not have its own independent source of income, it must rely on the "contributions" of its members and associated entities to pay the bill. And since it is extremely difficult, if not impossible, to devise a scheme for making these contributions which will have the continued approval of all of those involved, the ITU will undoubtedly continue to be plagued at intervals by requests for modifications. Under the circumstances, all that can be hoped for is that the response to such requests will not be a serious danger to the continued existence of the ITU itself.

TABLE 8.1
GROWTH OF ITU EXPENSES 1960-1981*

Year	Swiss francs
1960	10,270,423.50
1961	12,851,032.22
1962	14,708,688.68
1963	17,373,450.44
1964	16,885,328.33
1965	20,302,458.13
1966	23 306,590.00
1967	22,678,592.00
1968	24,150,328.00
1969	24,247,958.00
1970	27,132,089.00
1971	29,665,035.00
1972	32,000,151.00
1973	36,220,111.00
1974	45,444,140.00
1975	50,296,450.00
1976	56,716,549.00
1977	60,776,050.00
1978	59,912,132.00
1979	63,521,310.00
1980	64,749,900.00
1981	65,728,000.00 (Budget estimate)

*From ITU, *Report on the Activities of the International Telecommunication Union, 1961*, p. 54; *1962*, p. 56; *1963* p.62; *1964*, p.70; *1965*, p.70; *1966*, p.76; *1967*, p.68; *1968*, p. 74; *1969*, p.66; *1970*, p.78; *1971*, p. 102; *1972*, p. 96; *1973*, p. 106; *1974*, p. 128; *1975*, p. 114; *1976*, p. 102; *1977*, p. 122; *1978*, p. 110; *1979*, p. 126, ITU, Administrative Council, 35th Session, Geneva, May 1980, Doc. No. 5573-E, 30 May 1980, Resolution No. 849, p. 2; and 36th Session, Geneva, June 1981, Document No 5621-F, Addendum No 1 (27 April 1981), p. 72/3.

III

The Product

chapter 9

BASIC ARRANGEMENTS

The final product of the work of the various plenipotentiary and administrative conferences is contained in three basic documents: the International Telecommunication Convention; the Radio Regulations; and the Telegraph and the Telephone Regulations.[1]

The International Telecommunication Convention

The International Telecommunication Convention is the product of the efforts of the delegates to the intermittent plenipotentiary conferences. It is the ITU's basic document, in effect its constitution. The Convention gives the ITU its legal existence, establishes its structure, sets forth its purposes, defines its membership, fixes its relationship with the United Nations, and sets forth the primary regulations dealing with telecommunication in general and radiocommunication in particular.

The International Telecommunication Convention is divided into two parts, Basic Provisions and General Regulations. The Basic Provisions contain all of the fundamental rules and articles written in general terms, while the General Regulations contain complementary details of a more transitory nature. The present breakdown

---

[1]In the past there was also an Additional Radio Regulations which dealt with questions relating to rates and routing of messages in the aeronautical and maritime mobile services. The Additional Radio Regulations were created by the Washington International Radiotelegraph Conference of 1927, to collect all of the provisions of the then existing Radio Regulations dealing with matters which the United States was not prepared to accept. The transmission of radiotelegrams, as well as ordinary telegrams, was carried out in the U.S. by private enterprise which made provisions of the Radio Regulations difficult to apply. This permitted the United States to accept the general Radio Regulations without making a reservation. See Codding, *The International Telecommunication Union*, pp. 127-129. However, the delegates to WARC-79 felt that the Additional Radio Regulations no longer served a valid purpose and transferred all pertinent provisions back to the Radio Regulations.

between the two sections is the result of a decision of the 1965 Montreux Plenipotentiary that there might be some merit in creating a Constitution for the ITU which would contain all basic articles and which could only be modified by something more than a simple majority, as had been the case in the past.[2]

A special study group was created which worked from December 1967 to March 1969, to carry out the Montreux mandate. The report was presented to the 1973 Plenipotentiary at Malaga-Torremolinos. By 1973, however, the lesser developed countries were the majority and had second thoughts on giving a minority substantive veto rights concerning any fundamental changes in the structure or functions of the ITU. As a result, while it accepted the breakdown between the various provisions as suggested by the working group, the Malaga-Torremolinos conference refused to give the basic provisions the title constitution and decided to postpone any decision on whether to raise the majority needed to modify them until the next plenipotentiary.[3]

The Basic Provisions of the current convention begins with a preamble which reflects the paradox posed by any attempt to establish international regulations on matters of such great domestic import as telecommunication. "While fully recognizing the sovereign right of each country to regulate its communications," states this document, the delegations represented at Malaga-Torremolinos agree nevertheless "to establish this Convention which is the basic instrument of the International Telecommunication Union" and which establishes the rules which set limits on how countries of the world may use their telecommunications.[3] Subsequent materials in this section demonstrate the same basic dichotomy.

---

[2]See ITU, Study Group for the Preparation of a Draft Constitutional Charter, *Report of the Study Group Appointed to Prepare a Draft Constitutional Charter,* Document No. SGC/39-E, 5 March 1969, p. 2. A special study group was created which worked from December 1967 to March 1969, to carry out the Montreux mandate. The report was presented to the 1973 Plenipotentiary at Malaga-Torremolinos. By 1973, however, the developing countries were the majority and had second thoughts on giving a minority substantive veto rights concerning any fundamental changes in the structure or functions of the ITU. As a result, while it accepted the breakdown between the various provisions as suggested by the working group, the Malaga-Torremolinos conference refused to give the basic provisions the title "constitution" and decided to postpone any decision on whether to raise the majority needed to modify them until the next plenipotentiary.

[3]ITU *International Telecommunication Convention, Malaga-Torremolinos, 1973,* Preamble.

The first section of the Basic Provisions concerns the composition, purposes, and structure of the ITU. Membership in the ITU is open to all those countries contained in a list appended to the convention, members of the United Nations, and others who receive the approval of two-thirds of those already members.[4] The present membership of the ITU includes almost every independent country in existence, and as new countries achieve independence, they are admitted to membership without any difficulty whatsoever.[5]

Membership in the ITU involves an agreement to be bound by all the pertinent rules and regulations and the right to participate and vote in all conferences held under the auspices of the Union. Members are also eligible for election to the Administrative Council, to nominate candidates for any election to the permanent organs of the Union, and to vote in all consultations carried out by correspondence.

Considerable debate surrounded the decisions of the Montreux and the Malaga-Torremolinos Plenipotentiary Conferences to pass resolutions which denied South Africa the right to participate in ITU conferences, but such resolutions were passed and, although still a member, South Africa has not participated in an ITU conference since 1965.[6] In view of the almost "sovereign" nature of the power of the plenipotentiary conference, there is really nothing that can be done to overrule such a seemingly obvious breach of the convention as long as a majority of the participants are in agreement.

Although not stated in exactly these terms, the ITU has two inter-related purposes according to the Convention: (1) the maintenance of an efficient, world-wide telecommunication network, and (2) the constant upgrading of the technologies and procedures in that network. These generalities are defined more precisely as:

---

[4]Prior to the 1973 Plenipotentiary, there were provisions for both members and associate members and either could be composed of independent states or non-self governing territories. In 1973 the associate membership was abolished and the remaining non-self governing territories lost their membership.

[5]For a discussion of some of the earlier membership problems see Codding, *The International Telecommunication Union*, pp. 275-282 and 413-419.

[6]See, for instance, the 1973 debates which are found in ITU, *Minutes of the Plenipotentiary Conference of the International Telecommunication Union, Malaga-Torremolinos, 1973*, Geneva, 1974, pp. 142, 145-147, 154-177 and 180-227.

(a) To effect allocation of the radio frequency spectrum and registration of radio frequency assignments in order to avoid harmful interference between radio stations of different countries;

(b) To coordinate efforts to eliminate harmful interference between radio stations of different countries and to improve the use made of the radio frequency spectrum;

(c) To coordinate efforts with a view to harmonizing the development of telecommunication facilities, notably those using space techniques. . . ;

(d) To foster collaboration. . . with a view to the establishment of rates at levels as low as possible consistent with an efficient service and taking into account the necessity for maintaining independent financial administration of telecommunications on a sound basis;

(e) To foster the creation, development and improvement of telecommunication equipment and networks in developing countries by every means at its disposal, especially its participation in the appropriate programmes of the United Nations;

(f) To promote the adoption of measures for ensuring the safety of life through the cooperation of telecommunication services; and

(g) To undertake studies, make regulations, adopt resolutions, formulate recommendations and opinions, and collect and publish information concerning telecommunication matters.[7]

To accomplish these tasks, the ITU Convention authorizes a structure comprising the following organs:

1. The Plenipotentiary Conference, which is the supreme organ of the Union;

2. Administrative Conferences;

3. The Administrative Council;

4. The permanent organs of the Union, which are:

   a) the General Secretariat;
   b) the International Frequency Registration Board (IFRB);
   c) the International Radio Consultative Committee (CCIR);
   d) the International Telegraph and Telephone Consultative Committee (CCITT).[8]

---

[7]From ITU, *International Telecommunication Convention, Malaga-Torremolinos, 1973*, Art. 4.

[8]Ibid., Art. 5.

Also included is a Coordination Committee. Seven pages of text spell out their structure and functions. Since the substance of this material has already been considered, at this point a simple listing is all that is necessary.

The Convention also provides for a class-unit system for the payment of contributions by member states for the financing of the organization, gives the Union legal status in the territory of member states and designates the official languages (Chinese, English, French, Russian, and Spanish) and the working languages (English, French, and Spanish). In practice, this means that the final documents of conferences and official service documents are drawn up in all of the five official languages while all other documents are issued only in the three working languages. Further, interpretation into all five official languages is provided in all official ITU conferences and meetings, if needed.[9]

The next major section of the Convention contains the general rules concerning telecommunication and radiocommunication that the members of the Union agree to be bound by. The main emphasis is the maintenance of an efficient worldwide telecommunication network. Members of the ITU agree to provide the channels and other installations "necessary to carry on the rapid and uninterrupted exchange of international telecommunications," to maintain these channels and installations in operating conditions which keep abreast of scientific and technical progress, and to operate them using the "methods and procedures which practical operating experience has shown to be the best."[10]

Members recognize the right "of the public to correspond by means of the international service of public correspondence," secrecy of communication is pledged, and "all telecommunications concerning safety of life at sea, on land, in the air or in outer space, as well as epidemiological telecommunications of exceptional urgency of the World Health Organization" are given absolute priority. Government messages are given second priority and the gold franc is established as the monetary unit for the composition of tariffs and the settlement of accounts between administrations.[11]

---

[9]In case of dispute, the French version prevails. Ibid., Art 16.
[10]Ibid., Art 23.
[11]Ibid., Arts. 18, 22, 23, 25, 26, 28, 29, and 30.

While the emphasis in the General Regulations is on the smooth functioning of the international telecommunication network, the members of the ITU also reserve their right to make it less than perfect, if they so desire. States retain their right to stop the transmission of any private telecommunications "which may appear dangerous to the security of the State or contrary to their laws, to public order, or to decency." If a private telegram is thus censored, however, the country in question must notify the stoppage to the office of origin, except "when such notification may appear dangerous to the security of the State." As if this were not sufficient, the Convention permits members "to suspend the international telecommunication service for an indefinite time, either generally or only for certain relations and/or for certain kinds of correspondence, outgoing, incoming, or in transit" provided only that they notify such action to each of the other members of the Union through the Secretary-General. Finally, member governments refuse any responsibility to the users of the international telecommunication network, especially as concerns claims for damages.[12]

The section which contains the rules concerning radiocommunication is much more concise and to the point. Member states agree "to limit the number of radio frequencies and the spectrum space used to the minimum essential to provide in a satisfactory manner the necessary services" and to "apply the latest technical advances as soon as possible." Members agree that since "radio frequencies and the geostationary satellite orbit are limited natural resources" they should be used efficiently and economically in order to permit all countries to have "equitable access" to them "according to their needs and the technical facilities at their disposal."

Member states agree that all radio stations under their jurisdiction shall be operated in such a manner as to avoid harmful interference to other radio services or to stations of another country and to make certain that stations in the mobile radio services exchange communications with all other mobile stations without distinction as regards the radio system used by them. All radio stations are required to give absolute priority to distress messages, and member governments agree to take all necessary steps to prevent the

---

[12]Ibid., Arts. 19, 20 and 21.
[13]Ibid., Arts. 33-38.

transmission of false or deceptive distress, urgency, safety, or identification signals in their territories.

Even military stations are not completely exempt from international regulation. Although members of the ITU "retain their entire freedom" with regard to military radio installations of their army, naval, and air forces these installations must, "so far as possible," apply the appropriate provisions of the convention concerning distress messages and measures to prevent harmful interference and must apply the provisions of the Radio Regulations concerning types of emissions and frequencies to be used. Moreover, if military radio stations take part in the public correspondence service, they must comply with all of the appropriate regulations.[13]

The second part of the Convention, the General Regulations, spells out the details of the structure and functioning of the various organs of the ITU and establishes the general rules of procedure for the convening and conduct of conferences and meetings of the consultative committees. Again, since the procedures of conferences and meetings have been dealt with in some detail in earlier chapters, it will suffice at this point to take note of their presence in the convention and the fact that conferences and meetings have the power to adopt additional rules of procedure if they so desire.[14]

The Convention is completed by three Annexes, a list of member states, a list of definitions of certain terms used in the Convention and regulations, and a copy of the 1947 agreement between the ITU and the United Nations by which the ITU became a Specialized Agency.

Since plenipotentiary conferences must of necessity perform a number of tasks in addition to the revision of the Convention, the results of their deliberations are appended to the current convention for want of a better place to put them. These take the form of additional protocols, resolutions, recommendations, and opinions. The 1973 Convention, for instance, contains six additional protocols, forty-eight resolutions, five recommendations, and three opinions. As a general rule, the additional protocols contain elaborations or amplifications concerning provisions of the Convention, including in 1973, such things as the limits to the expenditures of the Union for

---

[14]Ibid., Arts. 14 and 53-82.

the period 1974 to 1979, an explanation of the procedure to be followed by members in choosing their contributory class, and the dates on which the newly elected Secretary-General and Deputy Secretary-General should take office. The resolutions are usually made up of directives to the other bodies of the Union. In the 1973 Convention seven dealt with the staff of the Secretariat, eight with finances, ten with various aspects of technical assistance, seven with future conferences and meetings, five concerned relations with the United Nations and its Specialized Agencies, and eleven were labeled "miscellaneous." The latter category included such things as a directive to the Secretariat to publish an annotated edition of the Convention, to study a structural question, and to establish May 17 as Telecommunications Day.

Interspersed among the resolutions are various political statements such as that concerning the admission of "Liberation Organizations" as observers to ITU conferences and meetings, the exclusion of Portugal and South Africa from conferences and meetings,[15] and a directive to the Administrative Council to expedite the procedures for the admission of Guinea-Bissau to membership in the ITU. Recommendations contain suggestions to member states and the UN on such matters as promoting the unrestricted transmission of news and making adjustments in the UN common pension system. Finally, the opinions section sets forth the position of the delegates to the plenipotentiary conference on such issues as the desirability of avoiding domestic taxes on international telecommunications, giving favorable treatment to developing countries in telecommunication relations, and encouraging the holding of telecommunication exhibitions organized under the auspices of the ITU.[16]

Finally, as an international treaty, the Convention must contain all of the housekeeping articles which are characteristic of such instruments.

The failure of some governments to carry out the necessary steps to ratify the ITU Convention has been a concern of the ITU almost from the beginning. In the ITU Convention, governments are requested to ratify the Covention as soon as possible. Since 1959, if

---

[15]The Portuguese exclusion was rescinded after the installation of a democratic government.

[16]Ibid., *Resolutions, Recommendations* and *Opinions*.

they have not done so within a period of two years, they lose their right to vote in ITU conferences and meetings and consultations by correspondence. The 1959 Geneva Plenipotentiary also discussed but rejected a proposal that failure to ratify would result in loss of membership (purging of the name of the country involved from the annexed membership list). Participation in the work of the ITU and adherence to the rules were considered more important than having such a severe punishment. The inclusion of the two year stipulation has improved matters somewhat, but there are still always a few countries which take a long time to get around to the ratification of the basic treaty.[17]

The 1973 Convention also contains a provision permitting a member to denounce the Convention by simple notification though diplomatic channels to the Secretary-General, an act which takes effect after one year,[18] and an article which provides that members "may" settle their disputes through ordinary diplomatic procedures. If such methods are not adopted a member "may" submit the dispute to arbitration according to a rather simple arbitration procedure provided for in the General Regulations section of the Convention. Neither of these two provisions has had a history of use. Finally, on January 1, 1975, the 1973 Convention comes into effect, at which time the old convention is abrogated.[19]

The ITU Convention also contains a Final Protocol. As permitted by the rules of international law, any state may take a reservation to a treaty. There have been many instances throughout the history of the ITU where governments have refused to be bound by various aspects of the conventions and regulations. The most obvious instance is the United States, which refused from the date of merger of the radiotelegraph and the telegraph organizations until recently

---

[17]As of 31 December 1979, for instance, five member countries which had signed the 1973 Convention had not yet submitted an instrument of ratification: Dominican Republic, Guatemala, Equatorial Guinea, Democratic Kampuchea and the Sudan. See ITU, *Report on the Activities of the International Telecommunication Union in 1979,* Geneva, 1980, Annex 1.

[18]Even the Russians did not withdraw from the ITU during the 1950s when they stopped participating in a number of other Specialized Agencies of the United Nations.

[19]See ITU, *International Telecommunications Convention, Malaga-Torremolinos, 1973,* Arts. 45-52 and 81.

to be bound by the Telegraph and Telephone Regulations, and from 1927 until it was eliminated in 1979, to be bound by the Additional Radio Regulations.[20]

The big increase in the use of reservations occurred in 1947 when it was decided at the instigation of the U.S.S.R. that the Regulations and the Convention were a package and that approval of the Convention also automatically bound the state in question to all of the Regulations. The 1947 Convention contains eleven reservations concerning the Telegraph Regulations, fifteen dealing with the Telephone Regulations, nine pertaining to the Additional Radio Regulations, and two to the Radio Regulations. In addition, there was a reservation by the U.S.S.R. concerning the portions of the Convention and the first annex which purported to refuse membership to Latvia, Lithuania, Estonia, and the People's Republic of Mongolia.[21]

During the 1952 Plenipotentiary Conference in Buenos Aires when the Cold War was at its height, a number of acrimonious debates took place in which the losers made remarks which implied that their governments might retaliate by withholding some of their contributions. For instance, when the U.S.S.R. delegation was denied its request to have Russian used in the conference on the same basis as the official working languages — English, French and Spanish — and paid for by all of the participants, not just those who agreed to its use, the delegates from the U.S.S.R. and eight other communist states made a declaration in the plenary assembly to the effect that their governments would pay only for the use of Russian and French.[22] When the time for reservations came around, a group of twenty-eight delegations reacted by introducing a reservation which declared that they would accept no consequence for reservations "resulting in an increase of their contributory share in the expenses of the Union."[23] In addition, another group of eleven

---

[20]For the origins and justification of the U.S. attitude towards the Additional Radio Regulations, see Codding, *The International Telecommunication Union*, pp. 127-129.

[21]See ITU, *International Telecommunication Convention, 1947, Final Protocol.*

[22]See ITU, *Documents de la Conference de Plenipotentiaries de l'Union Internationale des Telecommunications, Buenos Aires, 1952*, Geneva, 1953, pp. 64-67 and 118-126.

[23]ITU, *International Telecommunication Convention, Buenos Aires, 1952, Final Protocol*, XXXII.

introduced a reservation which claimed to reserve the right of their governments to take what action they deemed necessary if some other governments did not accept the decisions of the conference as regards the future work of the IFRB.[24]

A number of hotly contested debates also took place at the Geneva Plenipotentiary in 1959. By the end of the conference a number of delegations had decided that the formulas discovered in 1952 might again be appropriate and inserted both individual and group reservations to the same effect. More states adopted the practice at Montreux in 1965, and by the Malaga-Torremolinos Plenipotentiary in 1973 over 90 delegations were engaging in this somewhat pointless practice. The following reservation by Equatorial Guinea is typical:[25]

> The Delegation of the Republic of Equatorial Guinea reserves for its Government the right: 1.) not to accept any financial measure which might lead to an increase in its contributory share in defraying Union expenses; 2.) to take any action it deems necessary to protect its telecommunication services should any Member fail to observe the terms of the International Tele-communication Convention (Malaga-Torremolinos, 1973).

Other than this general statement of non-responsibility, the Final Protocols also include various political statements. The 1973 edition, for example, included a statement by a group of communist states complaining about the representation of Viet Nam, Korea, and Cambodia, along with a reply from those three states; a statement from a group of Arab states attacking Israel, along with a reply by Israel; and a complaint by Argentina that the Falkland Islands do not belong to the United Kingdom, along with a counter-reply by the United Kingdom. It is questionable whether the Final Protocols to ITU Conventions are worthwhile in light of their limited contribution to the smooth functioning of the international telecommunications network.

The exact relationship between the Convention and the Radio Regulations and the Telegraph and Telephone Regulations has

---

[24]Ibid., XXXVI.

[25]See ITU, *International Telecommunication Convention, Malaga-Torremolinos 1973, Final Protocol* LXIII.

been a controversial issue for the members of the ITU. While the 1932 Madrid Plenipotentiary decided that the administrative regulations were binding only on the members which agreed to be bound by them, the Atlantic City Plenipotentiary came to the conclusion that for the proper functioning of the international telecommunication network, it was necessary that all members should carry out all of the rules and regulations for all of the various telecommunication services. Thus, the Convention and Regulations should be considered a package and members should be bound by the Regulations in force at the time of ratification or accession to the Convention. This gave rise to numerous reservations to one or another of the regulations and a long debate at the 1952 Buenos Aires Plenipotentiary between those who felt that membership in the ITU carried the responsibility to carry out all of the Regulations and those who preferred a more flexible system. The 1952 Plenipotentiary came down on the side of the flexible approach and decided that members of the ITU had the obligation to approve and apply only one of the major Regulations. No change was made until 1965 when the present system was adopted by which both of the Regulations are considered to be an essential part of the Convention and all members are obliged to carry out the rules contained in each.[26]

The Radio Regulations

The basic arrangements set forth in the Convention are fleshed out in greater detail in the administrative regulations. The Radio Regulations is the final product of general world administrative radio conferences, such as WARC 79 and specialized administrative radio conferences. The Telegraph Regulations and the Telephone Regulations are the product of the periodic World Administrative Telegraph and Telephone Conferences, the latest of which was held in Geneva in 1973.

The most complicated of the three regulations since 1973 is the Radio Regulations. These regulations are divided into two parts: Part A - General Provisions, and Part B - Provisions Relating to Specific Services. General Provisions contain a number of sub-

---

[26]With the decision on the part of the United States that it can live with the Telegraph and Telephone Regulations after all, there should be no difficulties surrounding this provision in the future.

divisions, including terminology, technical characteristics of stations, measures against interference, administrative provisions for stations, and service documents. Probably the most important and complicated portion of this section is contained in Chapters III and IV, which deal with the allocation of the frequency spectrum to various "services," and the manner in which frequencies are notified and registered with the ITU's International Frequency Registration Board. This is the heart of the Regulations and the area which has always caused the most controversy since the Atlantic City Radio Conference of 1947.

Chapter III is a description of radio use throughout the world, which is based largely on historical accident and political negotiations but to some extent on the efficient use of frequencies. The Frequency Allocation Table in Chapter III allocates radio frequency bands from 9 kHz to 400 GHz to several dozen radio services throughout the world. Where the frequencies have worldwide propagation characteristics, an attempt is made to require the same usage worldwide, although in many instances frequency bands are shared among two or more users. Other bands are frequently broken down into regional usage and many of the regional bands allow a sharing among two or more users. In addition, individual states and groups of states are permitted during a radio conference to signal a deviant usage. These "footnotes" to the Frequency Allocation Table have increased tremendously over the years and must be taken into account if one is to know exactly the purpose to which any particular frequency band is being put. Delegates to conferences have been urged to use as few footnotes as possible, but for many reasons the practice has grown steadily.

At each radio conference, especially a general one such as WARC 79, there is a deluge of proposals to find frequency space for new uses and to make changes in the allocation table to recognize an increase in some uses and a decrease in others. In addition, advances in the state of the art make it feasible to extend the upper limits of the table. For example, WARC 79 extended the upper limit from 275 to 400 GHz.

Chapter IV of the Radio Regulations contains the instructions to administrations on the procedure by which they are to register radio frequency assignments 1) if the use to which it is put is capable of causing harmful interference to the communications of another country, 2) if the frequency is to be used for international radio-communication or, 3) if it is desired "to obtain international recognition of the use of the frequency,"[27] and the procedure required to be carried out before an administration introduces a new satellite communication service. This chapter also establishes the procedure by which the Registration Board records these notifications in the Master Frequency Register. The IFRB and the International Frequency List are discussed at some length elsewhere in the book. These procedures have become more and more involved over the years as conferences have continued to make changes and additions. WARC 79 was to attempt to consolidate the provisions of this chapter and to make them more intelligible to administrations, but due to the press of work was not able to carry out this task and, in fact, succeeded in making them even more complicated.[28]

Part B of the Radio Regulations concerns procedures for the staffing and operation of the various types of radio services, including the rules for licensing of stations and call signs to be used,[29] inspection of radio stations in certain services, operator's certificates, working conditions in the mobile services, and procedures to be used in working with other stations.

The Radio Regulations has a section on distress, alarm, urgency and safety messages, containing the familiar SOS and MAYDAY calls, terminology, rules for the assignment and use of frequencies, and the ever important measures to eliminate interference between radio stations.

The Regulations also include some forty-four appendices which supplement various provisions in the main document.

The Radio Regulations are contained in a Final Acts, the official final document of the World Administrative Radio Conference that adopts them. These Final Acts also contain numerous resolutions

---

[27]ITU, *Radio Regulations*, (1982), Art. 12.

[28]The existing procedures are so complicated that WARC 79 directed the IFRB to prepare a special handbook to explain them.

[29]It is in this section of the Radio Regulations that one finds the call sign letters attributed to the countries of the world and various international organizations, i.e., for the U.S. the familiar AAA—ALZ, KAA—KZZ, NAA—NZZ, and WAA-WZZ.

and recommendations dealing with radiocommunication, the convening of specialized administrative radio conferences, and directives to other bodies of the ITU such as the CCIR and the IFRB. When added to the appendices to the Radio Regulations, these resolutions and recommendations make up a document that is much larger than the Radio Regulations proper.

If the Convention is complicated by the tendency of administrations to make reservations asserting their rights to deviate from the rules that they have just agreed to, it is no more complicated in this respect than the Radio Regulations. First, the Final Acts of the 1979 World Administrative Conference starts with the following general statement:

> The delegates of the Members of the International Telecommunication Union represented at the World Administrative Radio Conference, Geneva, 1979, having signed these Final Acts, hereby declare that, should an administration make reservations concerning the application of one or more of the provisions of the Radio Regulations, Geneva, 1979, no other administration shall be obliged to observe that provision, or those provisions, in its relations with that particular administration.[30]

Second, the Final Acts contain at least eighty different reservations, most of which are individual, but twelve of which were sponsored by two or more delegations. There were three general types of reservation, the political, the "formula," and the technical. There were seventeen political reservations concerning Israel, the Pol Pot regime in Cambodia, and sovereignty over the Falkland Islands. There were also reservations and counter-reservations concerning frequencies for various Antarctic territorial claims and for the U.S. naval base at Guantanamo Bay, Cuba, which could be considered as political. The "formula" reservations, of which there were at least thirty-three, assure the world that the administration making the reservation will take "all action it deems necessary" to safeguard its interests "should any Member of the ITU fail in any way to comply with the Radio Regulations."[31]

---

[30]See ITU, *Final Acts of the World Administrative Radio Conference, Geneva, 1979*, Geneva, 1979, *Preamble.*

[31]Although the language varied, the Zambian reservation is a good example of this type. Ibid., *Final Protocol.*

Of more potential importance to some telecommunication uses were the reservations to various aspects of the technical decisions that had been made at WARC 79, especially those dealing with the Table of Frequency Allocations. There were thirty-nine such reservations in all, involving some sixty-three different countries. The country with the most such reservations was the United States.

The Telegraph and Telephone Regulations

In contrast to the Radio Regulations, the Telegraph Regulations and the Telephone Regulations are a model of simplicity and brevity. The Radio Regulations run to almost 1,000 pages but the Telegraph and Telephone Regulations combined barely fill 70 pages.[32]

The reason is the revolutionary 1973 decision to turn most standards making over to the CCITT to allow them to be more easily updated, and thus more responsive to the rapidly evolving technology. The great mass of the earlier Telegraph and Telephone Regulations was therefore eliminated and replaced by the following article (which is similar in both sets of regulations):

1. (1) The Telegraph Regulations lay down the general principles to be observed in the international telegraph service.
(2) In implementing the principles of the Regulations, Administrations (or recognized private operating agencies) should comply with the C.C.I.T.T. Recommendations, including any Instructions forming part of those Recommendations, on any matters not covered by the Regulations.

2. These Regulations shall apply regardless of the means of transmission used, so far as the Radio Regulations. . . do not provide otherwise.[33]

The two sets of regulations are separate and comprise two distinct documents. However, the format and content are almost identical, and the bulk of the following description of the Telegraph Regula-

---

[32]Excluding the pages devoted to a reproduction of the signatures of the delegates.

[33]ITU, *Final Acts of the World Administrative Telegraph and Telephone Conference* (Geneva, 1973), *Telegraph Regulations, Telephone Regulations,* Geneva, 1973, Art. 1.

tions also applies to the Telephone Regulations. The stated purpose of the Telegraph Regulations is to "lay down the general principles to be observed in the international telegraph service."[34]

As concerns the telegraph portion of the international telecommunication network, the technical stipulations of the past are encapsulated in two paragraphs:

1. The circuits and installations provided for in the international telegraph service shall be sufficient to meet all requirements of the service.

2. Administrations (or recognized private operating agencies) shall cooperate in the establishment, operation and maintenance of the circuits and installations used for the international telegraph service to ensure the best possible quality of service."[35]

The Telegraph and Telephone Regulations both contain a description of the services offered to users, a short section concerning operations, and a set of definitions. The largest portion of these two slim documents is made up of rules concerning tariffs and how to apply them. Both also have an annex dealing with the payment of the balances of accounts between administrations.

The Final Acts of the 1973 Administrative Telegraph and Telephone Conference, in which the most recent Telegraph Regulations and

---

[34]Ibid.

[35]Ibid., Art. 3. The Telephone Regulations are a little more detailed, as befits a service which is more complicated technologically and is still in the development stage in many areas of the world:

1. All administrations [or recognized private operating agencies] shall promote the provision of telephone service on a world-wide scale and shall endeavour to extend the international service to their national network.

2. Administrations [or recognized private operating agencies] shall designate the exchanges in the territory they serve which are to be regarded as international exchanges.

3. The circuits and installations provided for the international telephone service shall be sufficient to meet all requirements of the service.

4. Administrations [or recognized private operating agencies] shall cooperate in the establishment, operation and maintenance of the circuits and installations used for the international telephone service to ensure the best possible quality of service.

5. The Administrations [or recognized private operating agencies] shall determine by mutual agreement which routes are to be used. Ibid., Art. 3.

Telephone Regulations are found, also contain a series of resolutions, recommendations, and opinions dealing with telegraph, telephone, and related ITU matters. In contrast to the Final Acts of the 1979 World Administrative Radio Conference or even the 1973 Convention, there are only twelve. There were six resolutions, four dealing with telegraph matters, one concerned the telephone, and one dealt with the non-participation of South Africa.[36] Of the three recommendations, one dealt with the payment of balances of telegraph and telephone accounts, one with routing of outgoing telephone traffic and one with United Nations telephone calls "in exceptional circumstances."[37]

The 1973 Final Acts also contain a Final Protocol in which the delegates to the 1973 Administrative Telegraph and Telephone Conference could make political statements and reservations to the decisions of the conference. There were only seventeen in the 1973 Final Protocol, eight political statements and nine reservations, seven of which were general and two specific. The eight political statements were all concerned with South Africa and Viet Nam. The two specific reservations were the refusal of the United States to be bound by certain aspects of the Telegraph and Telephone Regulations. Of the seven general reservations, only three were similiar to those in the Convention, whereby administrations declare that they will take "all measures" they deem necessary to protect their interests if other administrations fail to observe the provisions of the regulations.[38] Three other delegations made the obvious statement that their governments reserved the right to accept or reject the results of the 1973 conference,[39] and the Mexican delegation declared that: "In signing the Telegraph and Telephone Regulations, Mexico reserves the right to apply the Recommendations of the C.C.I.T.T. insofar as they may solve problems of a world-wide international character and meet regional requirements."[40]

---

[36]The resolutions concerning the telegraph included: Instructions for the Operation of the International Public Telegram Service; Revised Terminal and Transit Rates for Telegrams; Telex Operation and Tariff Principles; and a listing of telegraph related documents that the General Secretariat was directed to publish. See Ibid., *Resolutions*.

[37]Ibid., *Recommendations*. The remainder were opinions dealing with franking privileges for delegates to ITU conferences and interim measures for the interpretation of affected radio regulations.

[38]Algeria, Somalia, and the Sudan.

[39]Jamaica, Libya and Roumania.

[40]Ibid., *Final Protocol* No. VI.

Conclusions

The major output of the periodic meetings of delegates from member governments in ITU plenipotentiary and administrative conferences is contained in the International Telecommunication Convention, the Radio Regulations, and the Telegraph Regulations and the Telephone Regulations.

The Convention contains the basic statement of the desirability of having an efficient world-wide telecommunication network, the need for cooperation between states to achieve it, and the organizational structure within which this cooperation can take place. The Convention also contains the fundamental rules dealing with telecommunications in general and radio in particular, which administrations must obey to make the international network effective and efficient. The details are fleshed out in the three regulations, in a great deal of detail in the Radio Regulations, and in a minimum of detail in the Telephone Regulations and the Telegraph Regulations.

Throughout all of these, there is a masterly blend of obligatory requirements and convenient escape clauses, which is due to the recognition of states that there are rules that must be followed if the international telecommunication system is to work while, at the same time, an attempt is made to give the appearance that in following these rules the states in question are not giving up any of their precious sovereignty. As long as states retain their basic nature, the dichotomy will remain. Because of the importance of telecommunication, the ITU seems to work, nevertheless.

TABLE 9.1
OUTLINE OF THE 1982 RADIO REGULATIONS

Part A. (General Provisions)

I   Terminology
II   (Technical Characteristics of Stations)
III   Frequencies
IV   Coordination, Notification and Registration of Frequencies.
V   International Frequency Registration Board
VI   Measures Against Interference. Tests
VII   Administrative Provisions for Stations
       (Service Documents)

Part B. (Provisions Relating to Specific Services)

VIII Provisions Relating to Groups of Services and to Specific
     Services and Stations
IX   Distress and Safety Communications
X   Aeronautical Mobile Service
XI   Maritime Mobile Service and Maritime Mobile-Satellite
       Service
XII   Land Mobile Service
XIII   (Entry into Force of the Radio Regulations)
       Appendices

chapter 10

STANDARDS

Introduction

The adoption of international telecommunication standards is the original *raison d'être* of the ITU. When the European administrations wished to interconnect their national telegraph systems to form a continental network, the International Telegraph Union was created to serve as a forum for adopting standards which would allow that to occur. It has been doing this ever since, setting standards for almost every aspect of telecommunication. This chapter will portray the range of standards established within the ITU.

It is interesting to note that a public international organization like the ITU isn't really necessary for the adoption of international technical standards. Indeed, several important private international standards making bodies have existed for many years. Some of the more prominent include the International Organization for Standardization (ISO), the International Electrotechnical Commission (IEC), the International Federation of Information Processing (IFIP), and the International Union of Radio Science (URSI). These organizations tend to serve the needs of manufacturers for rather detailed equipment standards, and for data communication standards, in areas where the ITU's jurisdiction has not been strongly asserted, or where its standards making process is too slow for rapid developments in the field. There is close coordination among all these organizations, however, including the ITU. Their representatives attend each other's meetings, their offices are often in close proximity to one another (for example, the ISO is located across the street from the ITU), and many of the participants in the meetings of these organizations are the same.

If at the founding of the International Telegraph Union, the national systems had been owned and operated by private corporations, it is possible that a public international organization might never have emerged. However, from the beginning, nearly all national governments asserted almost total control over communication. When those governments needed to coordinate their communication activities, public international organizations were created for that purpose.

Standards are found in all of the ITU's legislative enactments including the Convention, the Administrative Regulations, and the Consultative Committee Recommendations. Thus, all of the ITU bodies engaged in fashioning these instruments can be considered as standards making bodies, and the instrument in which the standard is found is superficially indicative of the force and effect of the provision. Those that are found in the Convention enjoy the greatest stature, and those in the Recommendations the lowest.

One must distinguish, however, between form and substance. The level of the document in which the standard is found is only an indicator of the willingness of nations to be bound, not of the actual force and effect. For example, the television signal standards found in the CCIR Recommendations have been so universally followed and incorporated in so much equipment that their stature exceeds that of any to be found in the Convention. This also indicates how closely the force and effect of standards is tied to economics. It is the nature of the standard, combined with the value of the equipment for systems in which that standard is embodied, which ultimately determines the standard's importance.

Standards are, in the abstract, both a help and a hindrance to the development of telecommunications. By obtaining common agreement on a standard, business and governments can be induced to implement new facilities and systems. On the other hand, the adoption of such a standard is usually based on existing technological assumptions which may well change. The investment of large amounts of capital in equipment based on a standard represents a significant impediment to altering that standard. In the real world of competition among the providers of telecommunication equipment and services, the adoption of standards also represents a way of creating market opportunities.[1] Fortunately, the increased use of computer controlled, dynamic telecommunication equipment is mitigating the inflexibilities caused by the adoption of standards.[2]

In the case of the CCIR Recommendations, the various provisions may enjoy an additional force and effect through two mechanisms. In some cases, the Recommendations may form the basis for the adoption of a provision in the Radio Regulations or in a regional agreement such as an allotment plan. This may be either implicit through the utilization of a recommendation during the legislative process; or explicit through reference in the text of the regulation. The second mechanism is through "judicial notice" of a CCIR Recommendation during the course of an International Frequency Registration Board rights vesting proceeding. Thus, in reaching a finding in a particular controversy, the board may rely upon a CCIR recommendation as determinative.

The preponderance of ITU standards are contained in three sets of documents: the Radio Regulations, the CCITT Book Applicable After the Seventh Plenary Assembly (1980) (also formally referred to as the Yellow Book), and the Recommendations and Reports of the CCIR, 1978 (informally referred to as the Green Books).[3] The Telegraph Regulations and the Telephone Regulations now contain little other than some arrangements concerning the settlement of accounts for international telecommunication. Essentially all the standards that were in these instruments are now embodied in the CCITT Yellow Book.

---

[1]See Rhonda J. Crane, *The Politics of International Standards: France and the Color TV War*. Norwood, New Jersey: Ablex Publishing, 1979.

[2]Until fairly recently, radiocommunication equipment has had to be designed for specific applications, often according to detailed standards applicable to those uses alone. This required manufacturers to build equipment that was physically different (even if only wired differently) to comply with standards which varied for different applications or parts of the world. Increasingly, new technology allows the physical equipment (i.e., the hardware) to be universally the same, tailoring its performance to specific applications by varying the programs (i.e, the software) in the microcomputer which controls all the components within the particular piece of equipment.

[3]The reader is reminded that the CCIR XIVth Plenary Assembly met at Geneva during February 1982 and adopted the Recommendations and Reports of the CCIR, 1982.

The mix of documents and formats are due to historical evolution and struggles for jurisdiction over various subjects. This has resulted in sets of different documents and approaches which are extraordinarily difficult to use and are often duplicative. The CCITT format is the best, due to its arrangement on the basis of subject matter in a kind of legislative code. Matters are made more confusing, and costs excessive (the new CCITT Yellow Book sells for approximately 1500 Swiss Francs or 750 U.S. dollars) by the practice of republishing all the documents every four years. The existing state of affairs seriously impairs the utility of ITU standards, especially for developing countries, which lack the luxury of being able to dedicate technical personnel to deal with these documents or the funds to purchase them. It is a matter ripe for attention at a future plenipotentiary conference.

The few standards that are contained in the International Tele-communication Convention, such as the establishment of the gold franc as the monetary unit for international tariffs, were treated in Chapter 9. Since that chapter also covered many of the standards that are found in the Radio Regulations, and Chapter 11 will deal with those pertaining to frequency management, the rest of this chapter will concentrate mainly on the standards of the CCITT and the CCIR.

The Standards of the CCITT

The most comprehensive and detailed standards are promulgated by the International Telegraph and Telephone Consultative Com-mittee at its plenary every four years. However, the mere establish-ment of a consensus within one of the CCITT study groups meeting between plenary sessions is generally sufficient to achieve *de facto* worldwide adoption of a new or altered standard.

The CCITT Yellow Book consists of a set of 30 soft cover books, containing a total of 5,570 pages.[4] These books contain the minutes, reports, resolutions, and opinions of the Seventh Plenary, the hundreds of recommendations of the CCITT, a list of terms and definitions, and an index. The questions under study by the CCITT are now published separately on a limited basis.

---

[4] Referred to as fascicles, and designated by a period and a number, e.g., Vol. II.2.

These Recommendations are organized on the basis of a comprehensive subject matter outline as shown in Table 10.2, below.[5]

This outline may be reorganized in subsequent years by administrative conferences, if not by the CCITT itself. For example, during the Seventh Plenary in 1980, the Italian Administration introduced a document urging reorganization of CCITT work into only four families:

1. Service, operation, and maintence
2. Network structure and switching
3. Network performance and transmission
4. Terminals and dedicated networks[6]

This new approach was considered and resulted in the adoption of an opinion recommending further study of the matter.[7]

A detailed analysis of all the CCITT Recommendations is beyond the scope of this book. However, it would be useful to review some of the more recent developments. The existing CCITT organization of this material — which relies on artificial or even non-existent divisions among telegraph, telephone, and data services — is used in this review.

Telegraph and Telematic Services

In light of recent telex switching advances, new standards have been adopted recently dealing with enhanced facilities for telex subscribers and the interconnection of private teleprinter exchanges with the telex network. For maritime mobile telex, a new set of standards has been prepared for such characteristics as ship station identification, service codes, quality objectives, and store and forward facilities. Recommendations have also been adopted to enhance the operation of telex terminals by the addition of display screens and certain automatic functions.

---

[5]See CCITT VIIth Plenary Assembly, *Yellow Book*, Vol. I, Resolution No. 5.

[6]*Comments on the Study Groups and Their Work Programmes,* CCITT VIIth Plenary Assembly Temp. Doc. No. 6-E/PLEN (1980).

[7]See *Draft Opinion X,* VIIth Plenary Assembly Temp. Doc. No. 26-E/COM A [subsequently adopted by the Plenary (see Minutes of the 7th Plenary Meeting, VIIth Plenary Assembly Temp. Doc. No. 83-E/PLEN, para. 8.)].

In the teletex and videotex area, although only draft "descriptions" of the services were prepared, standards were adopted for the technical characteristics of terminals.

A number of new standards were adopted for public data networks. These include a Provisional Recommendation on customer access to packet switched networks via other types of networks; multiplexing specifications for high-speed subscriber lines; common channel signalling for both circuit switched and packet switched networks; network routing principles and maintenance, and the responsibilities of subscribers and administrations as to administrative arrangements in the use of networks. In addition, a Recommendation on a 1200 bit/s full duplex modem was approved. Many of these recommendations were so important that expedited approval prior to the Seventh Plenary had been obtained by a special mail-ballot procedure.[8]

Telephone Transmission and Maintenance

In the field of transmission media, a new presentation of line specifications was adopted to account for digital transmission uses. In addition, the first standards on fiber optic cables were adopted. New standards dealing with echo suppressors and cancellers were prepared. It should be noted that these equipment standards are illustrative of the considerable specificity of CCITT Recommendations, which in some areas actually describe the characteristics of individual hardware.

Because of the rapid worldwide transition from analog to digital networks, many new standards have been recently prepared. New Recommendations concerning maintenance principles and quality standards with respect to errors and permissible slip on international digital connections were adopted. A whole series of Recommendations have been prepared on digital line sections and systems on cable at both hierarchical and non-hierarchical bit rates. Important Recommendations on interconnection of digital paths using different techniques and on interfaces were also adopted.[9]

---

[8]See ITU, *Report on the Activity of the CCITT Between the VIth and VIIth Plenary Assemblies,* VIIth Plenary Assembly, Doc. No. 71 (1980) p. 19.
[9]Ibid.

Telephone Switching, Signalling and Operation

Because of the transition to digital systems, this area of CCITT standards is the most active. One of the most far-reaching developments lies in the preparation of Signalling System No. 7.[10] It is a common channel system which can be universally applied, is optimized for operation on digital transmission networks between stored program control (SPC) exchanges, can be used for an integrated services network, and is also capable of handling network management and maintenance traffic. Work is presently being conducted during the 1981-1984 study period on the development of standards for local digital exchanges intended for new services. In conjunction with the introduction of digital exchanges, the CCITT has developed three computer languages: CHILL (a high-level programming language for SPC exchanges), MML (a man-machine language to facilitate the execution of operational and maintenance functions of SPC switching systems), and SDL (a specification and description language for the graphical presentation of specifications and the description of the internal logic of SPC exchanges.)[11]

A series of amendments have been made to the existing recommendations on the World Numbering Plan to provide for new country codes, access codes to the international network, and to specify the capacity of international registers. In the area of traffic engineering, a thorough revision of the existing standard for measuring and recording traffic was adopted.[12]

Tariff Principles

The CCITT promulgates many important standards for establishing tariff charges for the use of public telegram, telex, telephone, facsimile, and leased circuit services. As the committee itself notes: "[t]ariff problems are always thorny and difficult to solve on account of their immediate and substantial financial implications for Administrations and recognized private operating agencies [and] explains the great diversity of views sometimes expressed at meetings and the difficulties encountered in arriving at a general

---

[10]See Yellow Book, Rec. Nos. Q. 701-741.
[11]See Rec. No. Z.200 for CHILL, and Rec. Nos. Z.101 - 104, Z. 311- 341 for SDL. Recommendations for MML are still under consideration.
[12]See Rec. Nos. E.410 and E.500. See generally, ibid.

consensus . . . "[13] Particularly controversial in this international dialogue is the matter of private leased circuits. Such circuits have the capacity to allow the lessor to provide an extremely wide range of services, including the resale of portions of its capacity. A new Recommendation had been recently adopted, laying down general principles for regulating the use of private international telecommunication networks by specialized organizations. This matter is controversial because if private businesses (i.e., "specialized organizations" in CCITT jargon) obtain their own circuits, they could potentially compete with the existing monopolies which own and operate the public networks.

The Recomendations concerning tariff principles for data transmission on various kinds of networks have also been recently expanded and updated. New recommendations have also been prepared for public telegram and facsimile services.

A new Recommendation was adopted for the payment of international telecommunication accounting balances aimed at meeting current requirements in monetary transactions. This action was taken in view of the unacceptability of the gold franc for a large number of ITU members who belong to the International Monetary Fund. The "interim procedure" provides a choice between using IMF Special Drawing Rights (SDRs) and the unilateral fixing of a gold franc value. The procedure is "interim" owing to the existing specification of the gold franc in the International Telecommunication Convention.[14]

Standards for Protection of Line Facilities

In conjunction with other international organizations such as the International Conference on Large High Tension Electric Systems (CIRGE), the International Electrotechnical Commission (IEC), the International Union of Railways (UIC), and the International Union of Producers and Distributors of Electrical Energy (UNIPEDE), the CCITT has promulgated many standards designed to protect transmission line and cable facilities. For example, new recommendations concern the "joint use of trenches and tunnels for telecommunication and power cables" and "methods for estimating

---

[13]Ibid, p. 19.
[14]Ibid.

induced voltage from radio station emissions and for reducing interference."[15]

Standards Promulgated by the CCIR

The CCIR, in contrast to the CCITT, has no generic codification or subject matter organization for its material. In addition, it has seven different levels of documents upon which it acts: recommendations, reports, resolutions, opinions, questions, study programs, and decisions. All of this material is organized on the basis of study groups, which in turn are largely oriented around various radio "services." The matter is further complicated by the fact that each study group organizes its material in a different way. The only practical method of finding a provision on a particular subject is to consult the Alphabetical Index of Technical Terms found in Volume XIII-2.

The published results of the CCIR's work are found in the CCIR Green Books which consist of a set of fourteen soft cover books containing a total of 3,790 pages. It should be noted that the CCIR Green Books, in contrast to the CCITT Yellow Book, contain all levels of documentation in essentially five different sections of each book. The first section contains recommendations and reports (reports are defined by the CCIR to be nascent recommendations), first ordered by working group, and then by recommendation number; the second section contains questions and study programs ordered by question number; the third section contains decisions ordered by decision number; the fourth section contains resolutions ordered by resolution number; and the fifth sections contains opinions ordered by opinion number. To complicate the matter even further, numbers and letters are used in the text number to indicate various levels of revision of a question and the branch of a study program; and suffixes are added to indicate the study group responsible for a text.

Most of the CCIR Recommendations pertain to frequency management. Many are adjuncts to provisions found in the Radio Regulations, and were drafted incidental to them. In some cases a Green Book provision may be explicitly referenced in the text of a Radio Regulation provision; in other cases, the reference is implicit and must be determined by examining applicable CCIR standards.

---

[15]Ibid.

Often, a body of CCIR standards is developed during the years preceding an administrative radio conference to support the drafting of a provision in the Radio Regulations at the conference. These close ties between the Radio Regulations and the CCIR standards tend to blur the distinctions between the two. There are several reasons for this tendency. The most obvious is that the same individuals who participate in the CCIR activities also play leading roles in the administrative radio conference. Somewhat less obvious is the need which the CCIR fills for a continuing forum for the development of legislation during the year prior to an administrative conference. There is no other mechanism for doing this. Finally, the distinctions between the Regulations and Recommendations are relatively superficial — a fact recognized by the telegraph and telephone community within the ITU that essentially led to their integration.

Standards Dealing With Frequency Management and Propagation

The majority of CCIR standards fall into the category of frequency management and propagation. Indeed, Volumes I (Spectrum Utilization), V (Propagation in Non-ionized Media), and VI (Propagation in Ionized Media) deal exclusively with these matters. All of the other CCIR volumes of standards (except XII) also contain many frequency management and propagation standards which are particularly relevant to certain applications, i.e., space research and astronomy, fixed, mobile, broadcast, satellite, and frequency and time signal dissemination.

Standards of this category are generally the oldest in the CCIR repertoire. This derives, in part, from the origin of the CCIR, and contrasts to the standards of the CCITT. The creators of the CCIT and CCIF in the 1920s fashioned standards needed to make the telegraph and telephone networks operate. Those active in founding the CCIR were largely a collection of scientists who were more interested in developing radio theory and advancing the state of the art than they were in assisting the administrative radio conferences to manage the spectrum. In addition, as a result of the considerable antipathy of the United States, France, and Great Britain toward the mere existence of the CCIR, it dealt only with theoretical matters in the beginning.[16] Many of the matters treated in Volumes I, II, V,

---

[16]See Codding, *The International Telecommunication Union*, pp. 121-22.

and VI thus address topics of a scientific nature.[17] As the years passed, the CCIR role was expanded, and standards for frequency management and specific systems were added in large numbers. This historical evolution is reflected in the organization of CCIR standards today.

Standards for Radio Systems

Standards for radio transmission systems, specialized networks, and the operation and maintenance of facilities are found throughout Volumes II, III, IV, and VII to XII. In some cases, one can even find standards for wire transmission systems, which would otherwise be expected in the CCITT Yellow Book, tucked away in CCIR volumes.[18] This has occurred because so many of the standards are of a specialized nature that the ones for wire transmission systems are placed under the jurisdiction of the CCIR, although they don't logically belong there. It is interesting to note, however, that the converse has not occurred to any significant extent. Thus the standards for radio systems ancillary to the wire telecommunication networks are not found in the CCITT Yellow Book, only in CCIR Volumes III, IV, VIII, and IX. One of the more recent exceptions to this tendency has been the study of standards for teletex, a service provided by television broadcasting, within CCITT jurisdiction.[19] These anomalies arise largely out of the historical evolution of the domestic and international bureaucracies which address these matters. However, the evolution of an integrated services digital network is rapidly eliminating most of the classical distinctions among telecommunication standards, necessitating a more comprehensive approach to standards making.

Most of the current activity of the CCIR is devoted to developing standards associated with satellite radiocommunication systems and digital transmission. Ranging from the very general[20] to

---

[17]See, for example, *Radiocommunication Requirements for Systems to Search for extraterrestrial Life,* Question 17/2, Vol. II, p. 464.

[18]See Vol. XII (Transmission of Sound Broadcasting and Television Signals Over Long Distances).

[19]See CCIR Vol. XI, Rec. 802, Opinion 60.

[20]See, e.g., "A Forecast of Space Technology," Report 672, Vol. II, p. 1; "Characteristics of Some Typical Experimental and Operational Communication Satellite Systems," Report 207-4, Vol. IV, p. 10; "Technical Characteristics of Systems in the Fixed Satellite Service," Question 2-3/4, Vol. IV, p. 319; Digital Radio-Relay Systems, Question 12-3/9, Vol. IX, p. 243.

specific new applications,[21] much of this work stands at the forefront of the widespread public availability of new equipment utilizing these techniques.

Conclusion

The standards making function of the ITU remains very important. However, the standards discussed in this chapter (in contrast to the ones associated with frequency management in the Radio Regulations, discussed in Chapter Eleven) are largely not contentious. The significant growth in the numbers of recognized private operating agencies (RPOAs) and scientific and industrial organizations (SIOs) participating in the work of the Union may prove responsive to the new technological environment. The new participants may well bring a new perspective and greater expertise to ITU forums. They can directly participate in matters of importance to them, effect the streamlining of organizational procedures and structures, and, perhaps most importantly, expand the jurisdiction of the ITU to encompass integrated telecommunication and information systems. Such changes will be imperative if the ITU is to continue to play a central role in international standards making.

---

[21]See, e.g., "Digital Interface Characteristics Between Satellite and Terrestrial Networks," Report 707, Vol. IV, p. 98; "Multiple Access and Modulation Techniques in the Fixed Satellite Service," Report 708, Vol. IV, p. 109; "Technical Factors Influencing the Efficiency of Use of the Geostationary Satellite Orbit by Radiocommunication Satellites Sharing the Same Frequency Bands," Report 453-2, Vol. IV, p. 181 (a report of special relevance to the ITU 1985 Space WARC, discussed in Chapter 14); "Inaccuracy and Reliability of Frequency Standards and Reference Clocks," Decision 29, Vol. VII, p. 116 (this work has special relevance to syncronizing worldwide digital networks); "Digital Selective-Calling for Future Operational Requirements of the Maritime Mobile Service," Question 9-3/8, Vol. VIII, p. 613; "Standards for Television Systems Using Digital Modulation," Question 25-1/11, Vol. XI, p. 339; "Standards for Digital Systems for the Transmission of Television Signals Over Long Distances," Question 14/CMTT, Vol. XII, p. 190.

TABLE 10.1
Outline of CCITT Yellow Book*

Volume I   Minutes and Reports of the Seventh Plenary
             Resolutions and Opinions of the CCITT
             Table of Study Groups and Working Parties during
             1981-84
             Summary of Study Questions during 1981-84
             Recommendations on Organization of CCITT Work
             (Series A)**
             Recommendations on Writing Style (Series B)
             Recommendations on General Telecommunication
             Statistics (Series C)

II .1 General Tariff Principles (Series D)
  .2 International Telephone Service - Operation
     (E.100-300)
  .3 International Telephone Service (E.401-543)
  .4 Telegraph and Telematic Services Operations and
     Tariffs (Series F)

III .1 General Characteristics of International Telephone
       Connections and Circuits
       (G.101-171)
   .2 International Analogue Carrier Systems.
      Transmission Media and Characteristics
      (G.211-651)
   .3 Digital Networks - Transmission Systems and
      Multiplexing Equipments (G.701-939)
   .4 Line Transmission of Non-Telephone Signals.
      Transmission of Sound Program and Television
      Signals (Series H, J)

---

*See Resolution No. 5, CCITT VIIth Plenary Assembly, Vol. 1,
Yellow Book (1980).
**The CCITT Recommendations consist of a "series" letter prefix
which designates the subject matter (See Table 10.2), followed by a
period, followed by a number.

IV.1 Maintenance; General Priniples, International
     Telephone Circuits(M.10-769)
  .2 Maintenance; International Voice Frequency
     Telegraphy and Facsimile, International Leased
     Circuits (M.800-1235)
  .3 Maintenance; International Sound Programme and
     Television Transmission Circuits (Series N)
  .4 Specifications of Measuring Equipment (Series O)

V   Telephone Transmission Quality (Series P)

VI.1 General Recommendations on Telephone Switching
     and Signalling.
     Interface with the Maritime Service (Q.1-119)
  .2 Specifications of Signalling Systems Nos. 4 and 5
     (Q. 120-164)
  .3 Specification of Signalling System No. 6 (Q.251-295)
  4 Specifications of Signalling Systems R1 and R2
     (Q.310-480)
  .5 Digital Transit Exchanges for National and Inter-
     national Applications. Interworking of Signalling
     Systems (Q.501-699)
  .6 Specifications of Signalling System No. 7(Q.701-741)
  .7 Functional Specification and Description Language
     (SDL). Man-Machine Language (Z. 101-104:Z.311-341)
  .8 CCITT High Level Language (CHILL) (Z. 200)

VII.1 Telegraph Transmission and Switching (Series R,U)
   .2 Telegraph and Telematic Services Terminal
      Equipment (Series S,T)

VIII.1 Data Communication Over the Telephone Network
       (Series V)
    .2 Data Communication Networks; Services and
       Facilities, Interfaces (X.1-29)
    .3 Data Communication Networks; Transmission,
       Signalling and Switching Network Aspects,
       Maintenance, Administrative Arrangements
       (X.30-180)

IX   Protection Against Interference. Protection of Cable
     Sheaths and Poles (Series K,L)

X .1 Terms and Definitions
  .2 Index of the Yellow Book

TABLE 10.2
CCITT Series Code for Recommendations

A Organization of the work of the CCITT
B Means of expression (definitions, vocabulary, symbols, classi-
   fication)
C General telecommunication statistics
D General tariff principles
E Telephone operation, network management and traffic engineer-
   ing
F Telegraph operation and tariffs
G Transmission: lines, radio-relay systems, radiotelephone circuits
H Utilization of lines for telegraphy and phototelegraphy
J Radio and television program transmissions
K Protection against interference
L Protection against corrosion
M Maintenance of telephone circuits and carrier systems
N Maintenance for sound-programme and television transmissions
O Specification of measuring equipment
P Telephone transmission quality. Telephone installations and
   local line networks
Q Telephone switching and signalling
R Telegraph channels
S Alphabetical telegraph apparatus
T Facsimile telegraph apparatus
U Telegraph switching
V Data transmission
X New data networks
Z Programming languages for SPC exchanges

TABLE 10.3
Outline of the CCIR Green Books*

I   Spectrum Utilization and Monitoring (Study Group 1)
        1A - Efficient Spectrum Utilization
        1B - Classification and Designation of
             Emissions
        1C - Specification and Measurement of Emission
             Characteristics
        1D - Monitoring
        1E - Noise
        Questions and Study Programmes, Decisions,
        Resolutions and Opinions

II   Space Research and Radioastronomy (Study Group 2)
        2A - Subjects of General Interest
        2C - Space Research
        2D - Radioastronomy and Radar Astronomy·
        2E - Amateur Satellite Service
        Questions and Study Programs, Decisions, Resolutions
        and Opinions

III   Fixed Service at Frequencies Below About 30 MHz (Study
        Group 3)
        3A - Complete Radio Systems
        3B - Radiotelephony
        3C - Radiotelegraphy and Facsimile
        Questions and Study Programmes, Decisions,
        Resolutions and
        Opinions

IV   Fixed Service Using Communication Satellities (Study
        Group 4)
        4A - Definitions
        4B - Frequencies, Orbits and Systems
        4C - Baseband Characteristics
        4D - Methods of Modulation and Multiple Access
        4E - Characteristics and Maintenance of Earth Stations

---

*See Texts of the CCIR, Vol XIII, *Recommendations and Reports of
the CCIR, XIVth Plenary Assembly, Kyoto, 1978.* The reader is
again cautioned that this outline was revised at the CCIR XV
Plenary Assembly which met in February, 1982.

4F - Frequency Sharing Between Networks of Fixed
Satellite Service
4G - Frequency Sharing Betwen Networks of the Fixed
Satellite Service and Those of Other Space
Radiocommunication Systems
4H - Frequency Sharing with Terrestrial
Radiocommunication Systems
Questions and Study Programmes, Decisions Resolutions
and Opinions

V   Propagation in Non-Ionized Media (Study Group 5)
5A - Texts of General Interest
5B - Effects of the Ground
5C - Effects of the Atmosphere
5D - Aspects Relative to the Broadcasting and Mobile
Services
5E - Aspects Relative to the Terrestrial Fixed Service
5F - Aspects Relative to Space Telecommunication
Systems
5G - Propagation Factors in Interference
Questions and Study Programmes, Decisions,
Resolutions and Opinions

VI   Propagation in Ionized Media (Study Group 6)
6J - Ionospheric Characteristics and Propagation
6K - Operational Problems
6L - Factors Affecting System Design
6M - Natural and Man-Made Noise
6N - Field Strengths Above 1.6 MHz
6P - Field Strengths Below 1.6 MHz
Questions and Study Programes, Decisions, Resolutions
and Opinions

VII   Standard Frequencies and Time Signals (Study Group 7)
Recommendations and Reports
Questions and Study Programmes, Decisions,
Resolutions and Opinions

VIII   Mobile Services (Study Group 8)
8A - Land Mobile Service and Land Mobile-Satellite
Service
8B - Maritime Mobile Service: Telegraphy and Related
Subjects
8C - Maritime Mobile Service: Telephony and Related
Subjects

8D - Maritime Mobile Satellite Service: Interface and
Access
8E - Maritime Mobile Service; Radio and Related
Subjects
8F - Aeronautical Mobile Service and Miscellaneous
Questions
Questions and Study Programs, Decisions, Resolutions
and Opinions

IX   Fixed Service Using Radio-Relay Systems (Study
Group 9)
9A - Performance Objectives
9B - Radio-Frequency Channel Arrangements and
Spectrum Utilization
9C - Interconnection Characteristics
9D - Maintenance
9E - Radio-Relay Systems for Special Applications
9F - Frequency Sharing With Other Services
Questions and Study Programs, Decisions, Resolutions
and Opinions

X   Broadcasting Service (Sound) (Study Group 10)
10A -Amplitude-Modulation Sound Broadcasting in LF,
MF, and HF Bands
10B -Frequency-Modulation Sound Broadcasting in VHF
and UHF Bands
10C -Sound Broadcasting in the Tropical Zone
10D -Recording of Sound Programs
10E -Broadcasting Service (Sound) Using Satellites
Questions and Study Programmes, Decisions,
Resolutions and Opinions

XI   Broadcasting Service (Television) (Study Group 11)
11A -Characteristics of Systems for Monochrome and
Colour Television
11B -International Exchange of Television
Programs
11C -Picture Quality and the Parameters Affecting It
11D -Elements and Methods for Planning
11E -Television Systems Using Digital Modulation
11F -Recording of Video Programmes
11G -Broadcasting-Satellite Service (Television)
Questions and Study Programmes, Decisions,
Resolutions and Opinions

chapter 11

FREQUENCY MANAGEMENT

Since 1903 frequency management has been a major function of the ITU and the organized international telecommunication community. It presently accounts for most of the ITU's international conference activity and more than fifty percent of its fiscal budget. This chapter will describe the various regulatory and administrative mechanisms that comprise the ITU's frequency management activities, and depict the ways these mechanisms have evolved over the past seventy-nine years. Much of this material also falls within the scope of other chapters dealing with the decision-making structure of the ITU (Section II) and the product (Section III). However, the subject of frequency management was only peripherally treated in those chapters; it will be brought together for cohesive treatment here.

Introduction

The terms "frequency management" and "spectrum management" (which are used interchangeably) refer to the adoption and the administration of various arrangements necessary for the efficient and interference-free use of radio. These arrangements are of two major types: (1) technical and operational standards and (2) administrative mechanisms that give a particular radio station the right to operate free from harmful interference from others. Governments around the world have their own national arrangements for this. The necessary international arrangements are devised largely through the ITU, although many separate regional and bilateral arrangements also exist.

The Nature of Radio

Before discussing frequency management, it would be useful to review briefly the uses of radio and some of the characteristics of radio waves. The most significant use of radio is for "radiocommunication."[1] Radiocommunication is effected through a source of invisible electromagnetic energy (generally a transmitter), which conveys information (usually by perturbing the energy with a modulator which is part of the transmitter), and radiates this energy into the surrounding environment (usually by an antenna). At another location, an antenna intercepts some of this radiated energy, which is then processed to extract the information (a receiver). In most cases the information is sound, pictures, or computer data.

Sometimes, the information is obtained by the passage of the radio waves between the transmitter and receiver, as in radar.[2] In other cases, the "transmitter" may be some far off star, as in radio astronomy.[3] All radiocommunication is a variation of these basic techniques involving the use of an energy source to convey information to a receiver.

The energy that is radiated is often referred to as radio waves. It passes through the environment as an invisible electromagnetic field, rapidly changing in magnetic and electrical polarity at a fixed rate per second. This rate of change is known as the frequency of the radio wave and is an important part of managing the use of the radio spectrum.[4] The electromagnetic energy of different frequen-

---

[1] "Radiocommunication" is defined to mean "telecommunication [i.e., "any transmission, emission or reception of. . . intelligence of any nature. . .] by means of radio waves. " *Radio Regulations* (1982), Art. 1, para. nos. 4,5.

[2] This technique is generically referred to as "radiodetermination" which is defined as "determination of position, velocity and/or other characteristics of an object, or the obtaining of information relating to these parameters, by means of the propagation properties of radio waves." *Radio Regulations* (1982), Art. 1, para. no. 10.

[3] "Astronomy based on the reception of radio waves of cosmic origin." *Radio Regulations* (1982), Art. 1, para. no. 14.

[4] For example, a standard broadcast (A.M.) radio station may be assigned a frequency of one megahertz (i.e., radiating an unmodulated "carrier" wave oscillating at a rate of one million changes per second) on which to operate. The scientific unit of measure for rate of change is a Hertz, abbreviated Hz. and usually accompanied with a prefix (kilo or k = 1000, mega or M = 1 million, giga or G = 1000 million). See *Radio Regulations* (1982), Art. 2.

cies possesses markedly different characteristics. Frequencies between nearly zero hertz and 3,000 GHz are regarded as radio.[5] Above 3,000 GHz as the frequency increases, the energy is described as infrared, light, ultraviolet, x-rays, and cosmic rays. Altogether, this continuum of frequencies is called the electromagnetic spectrum, and the radio portion is the radio spectrum. Even within the radio spectrum, the frequency of the energy makes a considerable difference in how it is propagated through the environment. Relatively low frequencies tend to stay close to the ground and travel great distances; medium frequencies both travel along the ground and travel greater distances, especially at night, by reflecting off invisible ionized layers of molecules many kilometer above the earth (the ionosphere); high frequencies tend to travel great distances largely by reflection between these ionized layers and the ground, to go bouncing around the world (so-called shortwave radio); and still higher frequencies don't reflect off the ionosphere at all, but tend to travel in a straight line, as does light. Most of today's use of radio is confined to frequencies between 10 kilohertz and 40 gigahertz. However, this is constantly changing as higher and higher frequencies come into use. Indeed, light communication by lasers has already proceeded from the laboratory to operational use between satellites in space.

Managing Radio Use

As noted, frequency management consists of establishing: (1) technical and operational standards and (2) administrative mechanisms that give a particular radio station the right to operate free from harmful interference from others. The first function consists of establishing mandatory specifications concerning the way radio equipment should perform, the way it should be operated (such as in emergency situations), and which portions of the spectrum should be reserved for particular kinds of radio uses. The second function is usually accomplished by defining certain radio "services" such as broadcasting, mobile, and radiolocation, and setting forth a table of frequency allocations indicating which frequency bands are reserved for which services.[6] Since the 1920s much of the work of frequency

---

[5] See *Radio Regulations* (1982), Art. 1, para. no. 6.
[6] See *Radio Regulations* (1982), Art. 1 for the definition of the various services, and Art. 8 for the Table of Frequency Allocations.

management has involved the preparation of and amendment to domestic and international tables of allocations.

The second major management function, determining radio station rights, is not always easily accomplished to everyone's satisfaction. When radio channels become limited, whether domestically or internationally, and the possibility of stations interfering with each other becomes likely, some kind of administrative arrangements must be established for deciding which facility, which individual, or which national administration will operate in a given geographical area. In the domestic situation, a government agency simply devises and enforces the methods necessary to accomplish this. On the international level, the matter gets more complicated because every nation regards itself as absolutely sovereign, unwilling to be governed by the dictates of any other nation or an international organization. As a practical matter, however, the desire to maintain interference-free radiocommunication has led most nations to abide by the arrangements fashioned at conferences of the ITU.

The arrangements for performing this second function are often described in a shorthand way, especially by lawyers and economists, as vesting rights in the radio resource.[7] An analogy to property can be made, and one can even speak of "vesting parcels of the resource." There is some danger, however, of oversimplification in constructing an analogy of this type. Thus, terms must be carefully defined. As used in this book the following meanings are intended. The "right" which is vested (i.e., authorized or granted) consists of an obligation imposed on all others not to cause harmful interference to the grantee's communications occurring within the authorized boundaries of the radio resource. The "boundaries of the radio resource" consist of one or more of the following characteristics associated with the use of radio: frequency channel, modulation type, spatial volume (including geographical area of coverage or geostationary orbit position), and time. The definition of these boundaries are critical to the entire rights vesting process. They also

---

[7] See, e.g., Harvey J. Levin, *The Invisible Resource — Use and Regulation of the Radio Spectrum*. Baltimore: John Hopkins University Press, 1971; Harvey J. Levin, "Orbit and Spectrum Resource Strategies — Third World Demands," *Telecommunications Policy*, June 1981; David M. Leive, *International Telecommunications and International Law: The Regulation of the Radio Spectrum*, Chapter 4.

tend to make the "parcels" so unique that they have little, if any, economic value between states.

The ITU's rights vesting function is usually accomplished either through the adoption of a "plan," through bilateral coordination, or through a notice and recordation procedure.[8] This complex area is discussed in greater detail below.

The remaining material in this chapter will be presented in two parts: the first part will examine the basic tools of frequency management as an introduction to the second part, which will trace the historical evolution of these tools during the past seventy-eight years.

The Nature of Frequency Management Arrangements

As previously noted, the arrangements for frequency management fall into two very different classes: technical and operational standards, and rights vesting mechanisms. Although there are some interrelationships between the two, their nature and purpose, as well as historical evolution, is quite disparate.

Technical and Operational Standards

The ITU produces many different kinds of technical and operational standards. Those that relate to the use of the radio spectrum are considered here. They can be subdivided into three categories: emission characteristics, allocations, and operational practices.

Radio transmitting facilities cause invisible electromagnetic energy to be radiated (or "emitted") into the environment. Other kinds of electrical equipment which have different purposes, such as microwave ovens and electric power lines, also leak radio energy as a byproduct of their operation, and can cause interference to radiocommunication. Technical specification which control the amount and kind of energy emitted into the environment are known as emission characteristics. These are probably the most highly tech-

---

[8] See *Radio Regulations* (1982), Arts. 10-17 and Appendices 25-27, 30 for worldwide arrangements; regional arrangements are generally effected through the adoption of plans annexed to the final acts of regional conferences.

nical of all the frequency management arrangements. Drawing upon the principles of physics and referring to the basic physical standards maintained by various national and international bodies, these specifications spell out in precise detail the emission characteristics of radio transmitters and ISM (industrial, scientific and medical applications) equipment.[9]

Typical emission characteristics specified in the Regulations and Recommendations concern field intensity, power levels, modulation methods, occupied bandwidth, carrier precision and stability, coding techniques, and satellite station stability, among others.[10] There are hundreds of standards in this category adopted to facilitate intercommunication and use the radio spectrum efficiently. Particularly important provisions are placed in the Radio Regulations, others are placed in the CCIR Recommendations, and some of intermediate import are in the Recommendations but incorporated in the Regulations by reference.[11]

The second category of technical and operational standards are allocations. The word "allocation" is a term of art in frequency management, meaning:

Entry in the Table of Frequency Allocations of a given frequency band for the purpose of its use by one or more terrestrial or space radiocommunication services or the radio astronomy service under specified conditions.[12]

Several dozen uses of radio are defined, such as broadcasting, mobile, and radiolocation, and certain bands in the frequency spectrum are specified primarily (or in some cases permitted and secon-

---

[9]"Operation of equipment or appliances designed to generate and use locally produced radio-frequency energy for industrial, scientific, medical, domestic or similar purposes, excluding applications in the field of telecommunication." *Radio Regulations* (1982), Art. 1, para. 16.

[10] See e.g., *Radio Regulations* (1982), Arts. 4, 5, 27, 28 and Vol. I, *Recommendations and Reports of the CCIR*, Kyoto, 1978.

[11]See, e.g., *Radio Regulations* (1982), Art. 5, paras. 300, 302, 305.

[12]Ibid., Art. 1, para. 17.

dary)[13] for a particular use or uses.[14] The allocation scheme is complicated due to the placing, in many cases, of several radio services in the same band, by giving different levels of prioity to services, by making different allocations in defined ITU regions,[15] and by introducing footnotes which indicate how individual countries or groups of countries intend to deviate from the general allocation scheme for the region.

There are often important technical reasons for making most of these allocations.[16] There are also safety, marketing, and scientific considerations which dictate certain results.[17] However, the bases for many, if not most, of the allocations lie in domestic and international political-economics. Various user groups find the allocation process a convenient means of securing a guaranteed monopoly of use or an advantage for themselves. Intense lobbying usually accompanies domestic or international allocation proceedings.[18]

The third category of standard is operational practices. This category consists of miscellaneous collection of arrangements, such as: a requirement to use station call sign prefixes designating the nationality of the station, and special codes or signals for calling or

---

[13] The terms "primary," "permitted," and "secondary" are defined levels of service rights to specified bands. See ibid., Art. 8, paras. 413 - 425.

[14] For example, the band 535 through 1605 kHz is exclusively allocated throughout the world to broadcasting. This is the standard A.M. broadcast band with which most people are familiar.

[15] These regions consist roughly of: Region 1 (Europe, Africa and the U.S.S.R.), Region 2 (the Western Hemisphere), Region 3 (Asia, Australia and Oceana), African Broadcasting Area, European Broadcasting Area, European Maritime Area, and Tropical Zone. See ibid, paras. 393 - 411.

[16] For example, the use of radar is basically incompatible with most other services, and usually receives an exclusive allocation.

[17] For example, certain frequency channels are set aside for emergency warnings or messages; certain television channels are designated for a large common market; and certain radio frequencies emitted by natural phenomena are protected for radio astronomers.

[18] In some countries, such as the United States, the international allocation process offers private-sector user groups the opportunity to help fashion international results which control domestic regulations, free from procedural legal constraints which would otherwise be imposed in domestic proceedings. See A. Rutkowski, "United States Policy Making for the Public International Forums on Communication," *Syracuse J. of Intl. Law and Commerce. Vol. 8, No. 1, (Summer 1981.)*

warning stations. The preponderance of these standards are contained in the CCIR Recommendations.[19]

Rights Vesting

The second major frequency management function involves the process of vesting a right to be free from harmful interference to the radio stations of each nation. This is the process of "allotment." It should be noted that the ITU has recently adopted a somewhat restrictive definition of the term:

> Entry of a designated frequency channel in an agreed plan, adopted by a competent conference, for use by one or more administrations for a terrestrial or space radiocommunication service in one or more identified countries or geographical areas and under specified conditions.[20]

However, there is an important need for a simple term that means "the process of vesting a right to be free from harmful interference," and the word "allotment" in its general sense has such a meaning. Therefore, throughout this chapter, the word "allotment" will be used in this general sense rather than the more restrictive meaning given in the current Radio Regulations.

There are basically two ways to allot the radio resource. The first is generically referred to as *a posteriori.*.It is the case-by-case approach based on experience and demand, usually relying on coordination among the interested parties, or on a notice and recordation procedure. It is often derogatorily referred to as "first-come, first-served." The second is generically referred to as *a priori.* It is the principled approach, usually relying on a negotiated plan based on a general formula or criteria for seeking equity among all the parties.

Over the last seventy-five years, one or the other approach has been advocated and used by nearly all nations, both internally and internationally. On a domestic level, the *posteriori* approach is often coupled with limited grant periods and an adjudication procedure for deciding among competing applicants. This makes the approach more equitable. On the international level, this is almost impossible

---

[19] See, e.g., *Recommendations and Reports of the CCIR*, Kyoto, 1978, Vol. VIII.

[20] *Radio Regulations* (1982), Art. 1, para. 19.

because of sovereignty. Most nations have been unwilling to allow an international body to adjudicate whether they can or cannot use a radio channel or satellite position. The ITU's International Frequency Registration Board (IFRB) was originally conceived to be this kind of quasi-judicial body, but was never given that authority. The result has been that whenever radio channels or satellite positions have become limited and contentious, the usual recourse has been to an *a priori* method.

The *a posteriori/a priori* division of rights vesting methods is the most basic way of characterizing them. Other attributes are: the length of time the rights are vested, whether they are conditional or not, the degree of specificity of the resource boundaries, the alienability or transferability of the resource, the institutional mechanism for vesting the right, the procedures for obtaining the right, and the susequent adjudication allowed. These attributes allow, in theory, for a wide range of different *a posteriori* and *a priori* rights vesting methods. In practice, relatively few methods have been used: notice and recordation, bilateral coordination, and multilateral planning.

Without question, the subject of rights vesting methods has been and remains the most difficult and contentious matter considered within the ITU. The majority of time expended at administrative radio conferences over the past sixty years has been devoted to rights vesting arrangements. Arguments for and against either *a posteriori* or *a priori* methods have been made repeatedly over the decades. When the availability of the resource is perceived to be limited by the international community, a concern for equity arises, and the pursuit of *a priori* methods has been the inevitable result. In opposition to this tendency have stood the nations which have acquired large allotments of the resource under pre-existing *a posteriori* methods; those which are concerned that detailed specifications and inflexible constraints accompanying *a priori* methods unnecessarily impede the implementation of new technology; nations unwilling to negotiate away their sovereign rights in achieving a plan; or those which disdain the tendency of nations to make highly exaggerated claims for allotments at planning conferences.[21]

---

[21] 10th Meeting of Sub-Committee B, Minutes of the Meeting of 8 July 1921; "Statement Made by Mr. George E. Sterling, Alternate Chairman, United States Delegation, High Frequency Conference, at the Plenary Assembly, April 8, 1949," International High Frequency Broadcasting Conference, Mexico City, 1948/9, Doc. No. 928; Rutkowski, "Six-Ad Hoc Two: The Third World Speaks Its Mind," *Satellite Communications,* Vol. 4, No. 3 (March 1980) p. 22.

The Historical Evolution of International
Frequency Management Arrangements

Technical and Operational Standards, 1906 to 1919

The earliest international technical and operational standards for radiocommunications were prepared and adopted at the Berlin Radio Conference of 1906.[22] These standards were in the form of Service Regulations appended to the Radiotelegraph Convention. Several provisions were established that can be classified as emission characteristics, allocations, and operational practices. For instance, under normal conditions, shipboard stations were not allowed to exceed a transmitted power of one kilowatt, and were required to use "syntonized" systems.[23] The 1906 Berlin Regulations also initiated the concept of "service" and what was to evolve into the table of allocations. Two service categories were set forth: "general public service" and "long-range or other" services.[24] These services were to share the one allocated band.

The Service Regulations also contained several frequency management operational standards. For example, it was implied that all stations would have call letters and that the Bureau of the Telegraph Union "shall see to it that the same call letters for several wireless telegraph stations shall not be adopted."[25] In addition, the first code for signalling to all other stations was adopted, the distress signal SOS.

No significant changes were made to the frequency management provisions of the 1906 Regulations at the London Radiotelegraph Conferences in 1912. However, the wavelength of 1800 meters was specified as an additional communication channel for distant public correspondence in exceptional cases, and the "purity of emission" specification that the systems be "sytonized" was altered to a

[22] See Service Regulations Annexed to the International Wireless Telegraph Convention (Berlin, 1906).

[23] That is, equipment that possesses a tuned circuit. This requirement eventually evolved into various "purity of emission" specifications.

[24] The general public service was given an exclusive allocation of the band 600 to 1600 meters (187 to 500 kHz). Other services could be assigned to any other frequencies, although 300 meters (1000 kHz) was recognized for use by stations in the general public service. See Service Regulations Annexed to the International Wireless Telegraph Convention (Berlin, 1906).

[25] Ibid., Art. XXXVIII.

requirement that "waves sent out shall be as pure and little damped as possible" and that "the use of transmitting devices in which the waves sent out are obtained by means of sparks directly in the aerial shall not be authorized except in cases of distress."[26] The latter change was the first which addressed the economic aspects of frequency management. The outlawed devices represented old technology which utilized an inordinate amount of radio spectrum. Newer technology was more expensive, but utilized much less spectrum. These kinds of tradeoffs subsequently formed the basis for the solutions to battles subsequently foughts in the ITU over the phasing out of spark transmitters in 1927 and 1932, and double-sideband shortwave broadcasting in 1979.

1920 to 1926

The field of frequency management changed considerably after the First World War. The Inter-Allied Radiotelegraphic Commission developed extensive changes to the 1912 London provisions. It was the work of the Commission which produced many of the frequency management approaches that exist today. An entire article in the Regulations was drafted to elaborate in considerable detail the "power and waves of stations."[27] Many restrictions were proposed on the use of old spark equipment in light of the new "continuous-wave" transmitters that had become available. In addition, a report was introduced as a technical basis for certain regulations, which represented the same kind of function that the CCIR was to perform sixty years later.[28] The rudimentary allocations effected in the 1912 Regulations were expanded into a comprehensive scheme, the true forerunner of the table of allocations of today. The Commission suggested dividing radiotelegraphic services into four categories: mobile, fixed, military, and special.[29] Wavelengths were allocated among the four services for the entire radio spectrum from zero

---

[26] Service Regulations Affixed to the International Radiotelegraph Convention, London, 1912, Art. VII.

[27] The French Republic, Office of the Minister of War, Cipher Section, EU-F-GB-I Protocol of the 25 August 1919 [sic], Art. VII.

[28] Cf. [Report of the] Project for Regulation of Wave-lengths and Powers Attributed to Fixed Stations, [appended to] EU-F-GB-1 Protocol of the 25 August 1919; CCIR, *Technical Bases for the World Administrative Radio Conference 1979*,Geneva, 1978.

[29] See ibid.,Art. 2.

through 12,000 meters.[30] During the War, beacon and direction finding technology had rapidly progressed, and the draft regulations in the Protocol contained extensive specifications for the operation of stations employing this technology. The draft amendments also required the Berne Bureau to assign blocks of call signs to all countries.

It should be emphasized, however, that the work of the many post-war telecommunication bodies — the Inter-Allied Commission, the Preliminary World Conference on Electrical Communications at Washington in 1920, the domestic commissions that met during 1920-21 to study the work of the Preliminary Conference, and the 1921 Paris Technical Committee — was not adopted per se by the international community. The work, however, was eventually edited by the Berne Bureau and became part of the proposals to the 1927 International Radiotelegraph Conference at Washington which actually adopted the new international regulations.

1927 to 1947

By 1927, many countries (especially the United States) had changed their attitudes about detailed international technical regulations. The technology was changing rapidly, and the attitude toward international institutions and arrangements changed from enthusiasm to hostility. This change was reflected in many areas of the 1927 Radiotelegraph Convention, including the technical and operational standards for frequency management.

All of the detailed specifications on emission characteristics worked out by the Inter-Allied Copmmission were distilled into several general statements:

Waves emitted by a station must be maintained upon the authorized frequency, as exactly as the state of the art permits, and their radiation must also be as free as practicable from all emissions not essential to the type of communication carried on.

The interested administrations shall fix the tolerance allowed between the mean frequency of emissions and the recorded

---

[30] See ibid., Table I, Allocation of Waves, It was not until the meeting of the Paris Technical Committee on Radio-communications in 1921, that a conversion to frequency rather than wavelength was made to denote the boundaries of service bands.

frequency; they shall endeavor to take advantage of technical improvements progressively to reduce this tolerance.

The width of a frequency band occupied by the emission of a station must be reasonably consistent with good current engineering practice for the type of communication involved.[31]

One of the most contentious frequency management issues at the conference involved the use of spark transmitters (emitters of so-called Type B waves). Although the international community may not have known what kind of emission specifications it wanted to apply in light of changing technology, it did know what it wanted to eliminate — spark transmitters.[32] This phasing out of spectrum inefficient equipment represented an important spectrum management consideration — the trade-off of spectrum value versus equipment value. In this case, spark, equipment used such a large bandwidth, that although many fought to retain it, consensus was reached concerning its eventual removal from use.

The 1927 Regulations promulgated the first comprehensive allocation scheme. Seven services (fixed, mobile, maritime mobile, broadcasting radio beacon, air mobile, and amateur) were given exclusive or shared use of various frequency bands between 10 kHz and 40 MHz.[33] The Table was to grow dramatically over the years and become the major focus of general administrative radio conferences. In some instances, particularly in the early decades of radio, there was a need for uniformity in using certain radio bands for prescribed uses. Important technical, operational, and economic reasons dictated various divisions. However, there were also very significant political considerations involved. It became an important vehicle by which radio industry groups laid claim to blocks of a valuable natural resource. Many a government official was lobbied

---

[31] See General Regulation Annexed to the International Radiotelegraph Convention, Washington 1927, Art. 4.

[32] Spark transmitters operating below 375 kHz were banned after 1 January 1930, except existing land stations; no new spark installations were to be made in ships or aircraft after the same date; no such installations were to be made in land or fixed stations effective immediately; no spark transmissions were to be made by fixed or land stations after 1 January 1935; and use was forbidden altogether after 1 January 1940. Ibid. Art. 5.

[33] Ibid, Art. 5. Article 5 has remained the heart of the *Radio Regulations*. In 1979 the Table was redesignated as Art. 8.

and cajoled over the years to shift band-edge frequencies a few kilohertz or megahertz one way or the other to obtain an advantage for broadcasters, the maritime industry, or common carriers. Because of this contention, global agreement was difficult. Even in this first adopted table of frequency allocations, distinctions were made to allow different uses in different regions of the world. In one of the low-frequency bands, one for other regions.

Several frequency management operational standards were adopted. A table and specification of call signals for every country was established.[34] Numerous "calling and listening waves" were specified[35] and standards for the format of automatic alarm, beacon, and compass signals were adopted.[36]

This period thus began with the formal adoption of international frequency management technical standards and approaches developed after World War I, which largely exist today. For the next fifty years, various provisions would be adjusted to reflect the improved state of the art of equipment and the advent of new services, but the basic standards would not be changed.

Coming only five years after the Washington conference, the Madrid Radiotelegraph Conference of 1932 did not make significant changes to the frequency management standards. Although the Table of Frequency Allocations was further subdivided in some bands, for the most part, the "General Regulations Annexed to the International Radiotelegraph Convention" simply became the "General Radio Regulations Annexed to the International Telecommunication Convention." The International Radio Conference convened at Cairo in 1938 further subdivided and rearranged the Table of Frequency Allocations, but did not otherwise make substantive changes to the standards.

The International Radio Consultative Committee (CCIR), which was created at the 1927 Washington Conference, did begin to meet during this period. Its study program reports and recommendations

---

[34] Ibid, Art. 14.
[35] Ibid, Art. 19.
[36] Ibid, Art. 21.

constituted additional standards which supported and amplified many of the more general provisions found in the Radio Regulations.

1947 to the Present

To a large extent, the Atlantic City Radiocommunication Conference in 1947 continued the standards making process begun several decades earlier. It did, however, change many significant details.

Probably the most important alteration of standards at the 1947 conference involved the Table of Frequency Allocations. The Second World War had produced radar and similar radiodetermination systems that needed to be accomodated in the Table. In addition, the advancing technology had pushed the upper boundary of the Table to 10,500 MHz. New services were contending for allocations which produced further fragmentation of the Table. This also resulted in a new arrangement which subdivided the world into three numbered regions: (1) Europe, U.S.S.R., and Africa; (2) North and South America; (3) Asia, Australia, and Oceana.[37]

Additional conferences would be held in the ensuing years to continue the process of modifying these basic provisions. The most significant alterations occurred in 1971 with the definition of many new satellite radiocommunication services, and their inclusion in the Table. At the WARC 79, the Table was even more fragmented through the extensive use of footnotes indicating special allocation for individual nations or groups of nations.

Today, the ITU's documents recognize thirty-seven distinct radio-communication services [38] and set forth numerous kinds of standards for them both in the Radio Regulations and the CCIR Reports and Recommendations. Major categories include: frequency bands allocated to the services in the radio spectrum between 9 kHz and 275 GHz (Article 8); technical characteristics of stations (Article 5, Appendices 7 & 8); measures against interference (Article 18-22); administrative provisions for stations (Articles 23-25); and a host of detailed provisions applicable to specific services (Article 27-68).[39]

---

[37]See *Radio Regulations* (Atlantic City, 1947), Art. 5.

[38]*Radio Regulations* (1982), Section III, Art. 1.

[39] See *Radio Regulations* (1982).

The Reports and Recommendations of the CCIR possess somewhat different categories which include; general spectrum management (Volume I), radio propagation (Volumes V and VI), and detailed provisions applicable to specific services (Volumes II-IV, VII-XI).[40]

During the 1980s, numerous administrative radio conferences and CCIR plenary assemblies will study and alter this body of standards further. Increasingly rapid technological changes will place great strains on this process. Because standards can serve both as a catalyst and a brake on innovation, it will require careful analysis to weigh both the advantages and disadvantages of specific provisions. In addition, the mechanisms for adopting these standards may need to be streamlined in order to rapidly adopt needed new standards, and to get rid of the unneeded old ones.

Rights Vesting Arrangements
1903-1911

International arrangements for vesting rights in the radio resource can be traced back to the first international agreement relating to the use of radio. Six of the eight countries attending the Berlin Preliminary Radio Conference of 1903 agreed to a Final Protocol which established the right that

> Wireless telegraph stations should operate, as far as possible, in such a manner as not to interfere with the working of other stations.[41]

A second provision established a notification process.

> States shall publish all technical information of a nature to facilitate and accelerate communications.[42]

These two provisions formed the initial, basic approach which governments used to implicitly establish their respective rights to interference-free use of radio stations.

---

[40] See *Recommendations and Reports of the CCIR*, 1978.
[41] Postal Department of the German Empire, Preliminary Conference Concerning Wireless Telegraphy, 4-13 August 1903, Final Protocol, Article IV.
[42] Ibid., Art. I.2.

Three years later, at the Berlin Radio Conference, which produced the first convention and regulations for radio, the two provisions in the 1903 Final Protocol were made more definitive by imposing an affirmative obligation associated with the right that

> The working of the wireless telegraph stations shall be organized as far as possible in such manner as not to disturb the service of other wireless stations.[43]

It also elaborated the details of an administrative process for notification:

> Parties shall notify one another of the names of coastal stations and stations on shipboard . . . as specified in the Regulations.[44]

In the annexed Regulations, it was specified that the "International Bureau [of the International Telegraph Union in Berne] shall be charged with drawing up a list of wireless telegraph stations . . ."[45] This list was first published as The Frequency List of Radioelectricity Stations (also known as the Berne List), and later as the International Frequency List (IFL). The same basic features still exist as elements of *a posteriori* notification and recordation process administered by the ITU.

The information notified to the Bureau was sufficiently descriptive to stake out a portion of the radio resource.[46] Once notified of this information, other stations were obliged not to cause harmful interference to the station which had acquired the right to communicate within the specified boundaries.

1912 to 1926

Six years later at the London Radiotelegraph Conference of 1912, the Convention and Regulations were amended and readopted. No

---

[43] International Wireless Telegraph Convention Concluded Between Germany, The United States of America, Argentina, Austria, Hungary, Belgium, Brazil, Bulgaria, Chile, Denmark, Spain, France, Great Britain, Greece, Italy, Japan, Mexico, Monaco, Norway, The Netherlands, Persia, Portugal, Roumania, Russia, Sweden, Turkey, and Uruguay (Berlin, 1906), Art. 8.

[44] Ibid., Art. 6.

[45] Service Regulations Annexed to the International Wireless Telegraph Convention (Berlin, 1903), Art. IV.

[46] These included: "normal wavelength" (i.e., frequency channel), "wireless system" (i.e., modulation type), "location" and "range" (i.e., spatial volume), and "hours of operation" (i.e., time).

alterations were made to the allotment provisions other than to replace the word "wireless" with "radio."[47] Because of the advent of the First World War and a growing antipathy toward international organization during the 1920s, the provisions of the London Convention were to remain in force for fifteen years.

During the years after the 1912 London conference, radio technology and its use changed dramatically, spurred primarily by the First World War. In addition, the war initially produced a world climate receptive to international arrangements and organizations such as the League of Nations. It was not surprising, therefore, to find the Allied Powers meeting to consider a dramatic revision to the 1912 London Convention by integrating it with the Telegraph Convention to create a Universal Electrical Communications Union which would, among other things, allot radio frequencies equitably among the nations of the world. Much of this effort was led by military technical advisers in the U.S. and France. The period is interesting and important not only because it fostered the first meetings to consider the comprehensive international regulation of the radio spectrum, but because it produced a remarkably well developed draft scheme for an equitable *a priori* allotment of the radio resource, including the possible trading of rights.

A series of meetings of the Inter-Allied Radiotelegraphic Commission at Paris during 1919, produced the "EU-F-GB-I Radio Protocol of the 25 August 1919." Soon after, the United States established a Department of Commerce Radio Conference Committee to consider the Protocol as the basis for a World Conference on Electrical Communications. That committee developed an allotment scheme which subsequently became the U.S. proposal to the Preliminary Conference at Washington in 1920. It was in large measure included in the draft Convention produced by that conference:

1) An International Technical Commission shall be. . . constituted for the purpose of assigning wave lengths. . .

2) [T]he party to which the said band of wave lengths is designated shall have the exclusive right under this assignment to transmit with said limits of wave lengths in any part of the world.

---

[47] See *International Radiotelegraph Convention* (London, 1912), Arts. 6 and 8.

When a transmitting station. . . has acquired recognition of a "right of way in the ether," this implies that no radiation shall be permitted in any part of the world which would interfere with the proper reception of its signals in any part of the world without the specific consent of the nation involved; and the assignment to a nation of any wave length by the International Communications Conference, or by international agreement upon the advice of the International Radio Technical Committee, shall carry the right for that nation to use, or permit the use of, and control the use of the assigned wave length in any part of the world, subject only to agreement of the local National Radio Administration involved.

In the allocation of these wave lengths to any nation, that nation shall have the right to make use of the same wave length or wave lengths for several systems of communication. . . The object of this provision is to encourage development of the art along these directions inasmuch as the limited number of available wave lengths should be used as effectively as possible to carry the greatest volume of radio communication.

3) The distribution of bands among nations would be in accordance with the proposition of General Ferrie of France and would be done on the basis of national population and number of remote colonies.

4) The bands allotted to a nation may be ceded to other nations provided the International Conference so authorized.[48]

An extensive draft allotment plan was also set forth.

In October and December 1920, a multilateral meeting referred to as the Preliminary International Conference on Electrical Communications was held for the purpose of preparing a draft convention and regulations for a proposed Universal Electrical Communication Union. This conference essentially adopted the rights vesting proposals of the U.S. Dept. of Commerce Committee, and set forth rather explicit provisions dealing with "tenure," "application," and "allocation of prior rights."[49] The allotment formula, drafted by General Ferrie of France at the meetings of the Inter-Allied Commission in 1919, was expanded to include three "coefficients of distribution to decide the maximum number of waves that may be

---

[48] See EU-F-GB-I Radio Protocol [Report], Washington, Government Printing Office (1920), pp. 88-98.
[49] See Universal Electrical Communications Union, *Draft Convention and Regulations (1920)*, p. 89.

allowed to any one administration."[50] These included: "(a) population, (b) the extent and importance of the colonies, dominions,
protectorates, etc., associated with the administrations and the
needs of communications with or between the same, and (c) the
extent of commerce and the need for international communications."[51]

The coefficients of distribution of all the different administrations
shall be added together and the total thus obtained will be taken
as equivalent to the total values of all the waves available for
distribution between 3050 and 30000 [meters]. Each administration shall be allocated a total width of wave band proportionate to
their coefficients of distribution.

If, in any case, the requests for wave bands received from any
administration are less than the total width of wave band
allowed, the surplus will be available for distribution among the
other administrations.[52]

The scene then shifts to Paris where the Technical Committee on
International Radio-Communication met for three months during
the summer of 1921. This committee was set up by the Washington
Preliminary Conference to consider several unsolved questions,
among them the equitable allotment question. After debating the
matter for hundreds of hours during the course of more than two
dozen sessions, it became clear that no formula for an equitable
distribution of wavelengths could be devised. In the end, the
committee could only suggest ". . . that some equitable distribution
of frequencies. . . should be made by the next International Conference, upon some basis which will combine suitable political
principles, to be determined at that Conference, with the technical
principles suggested by the present Technical Committee. . . "[53]

During the years after 1921, the spirit of internationalism began to
wane. In the United States, the notion of creating a new public
organization for telecommunications was defeated by a rather

---

[50] Ibid.
[51] Ibid.
[52] Ibid.
[53] [Report of the] Technical Committee on International Radio-Communication, p. 31.

strong sentiment against the League of Nations.[54] Because the U.S. was a major supporter for the *a priori* concept under the leadership of those surrounding President Wilson, lack of that support effectively terminated the matter. In addition, it became clear during the course of the almost endless debates at the Paris Technical Committee that worldwide agreement would be extremely difficult, if not impossible, to achieve.

1927 to 1947

It was not surprising that the 1927 Washington International Radio-telegraph Conference adopted no dramatic new mechanism for vesting rights. The matter, however, was a source of considerable controversy,[55] resulting in an attempt to explain the right involved in greater detail and to establish an arbitration procedure. The new provisions read:

> All stations, whatever their purpose, must, so far as practicable, be established and operated so as not to interfere with the radio communications or services of other contracting Governments and of individuals or of private enterprises authorized by these contracting Governments to carry on public radio communication service.

> The contracting Governments shall notify each other, through the intermediary of the International Bureau of the Telegraph Union, of the names of stations open to the international service of public correspondence and of stations carrying on special services covered by the Regulations.[56]

This basic scheme in the Convention was supplemented by the following detail added to the General [Radio] Regulations which

---

[54] For example, a "Memorandum of Radio Corporation of America with Reference to the Proposed Universal Electrical Communications Union" was printed in the form of a bound pamphlet in 1921 and widely distributed. In strong terms it stated ". . . it is the principle which underlies the proposed Union that is wrong. . . [i]t follows the principle of the League of Nations. . . with its attendant uncertainties of character and extent of the obligations." Ibid., p. 13.

[55] See John D. Tomlinson, "The International Control of Radiocommunications," Thesis No. 41, University of Geneva (1938) [also published by J.W. Edwards, Ann Arbor, Michigan (1945)] p. 142.

[56] *International Radiotelegraph Convention* (Washington, 1927) Arts. 8, 6.

made clear the "right of priority" (a term used extensively at that time):

The frequencies assigned by Administrations to all new fixed land or radio broadcasting stations which they may have authorized or of which they may have undertaken the installation must be chosen in such a manner as to prevent so far as practicable interference with international services carried on by existing stations the frequencies of which have already been notified to the International Bureau.

The interested Governments shall agree, in case of need, upon the determination of the waves to be assigned to the stations in question as well as upon the conditions for the use of waves so assigned. If no arrangement intended to eliminate interference can be arrived at, the provisions of Article 18 of the Convention [concerning arbitration] may be applied.

Each administration shall promptly advise the International Bureau when it decides upon, or authorizes, the establishment of a radio communication station, the operation of which necessitates the assignment for its regular service, of a particular frequency below 37.5 kc/s in the case where the use of this frequency might cause international interference over broad areas. This notice must reach the International Bureau four months prior to the construction of the station contemplated in order to dispose of objections which any of the Administrations might raise against the adoption of the proposed frequency.[57]

At the end of the 1927 Washington conference, the chairman of the conference, U.S. Secretary of Commerce Herbert Hoover, made vague reference to the earlier struggle in his closing remarks:

[A]t one time it was proposed that the use of these channels through the ether should be divided among the different countries of the world. That would have been equal to an assignment of different lanes across the seas upon which the vessels of a particular nationality should travel exclusively. It soon became appar-

---

[57] *General Regulations Annexed to the International Radiotelegraph Convention* (Washington, 1927), Art. 5.

ent that this solution would lead only to greater confusion, to international jealousies, and to injustice.[58]

The "Hoover solution" adopted by the conference was to maintain the status quo by making minor changes and readopting the *a posteriori* mechanism that had existed since 1903. The struggle between the years 1919 and 1927 to develop world arrangements for vesting rights is illustrative of some of the inherent tensions and conflicts which have manifested themselves over the last sixty years. Virtually the same struggle was repeated after the Second World War, and is occurring today.

It should be emphasized, however, that the inability to implement global *a priori* vesting mechanisms did not mean such mechanisms were not used. Indeed, the problems of interference among European nations in the mid-20s led to the adoption of the first allotment plan as early as 1926. Similar regional *a priori* mechanisms have existed around the globe ever since.

In 1925 the British Broadcasting Company (BBC) called many European administrations together to establish a means of cooperating in the use of radio broadcasting. The meeting produced the International Broadcasting Union (UIR), and one of its main tasks was to consider the problem of mutual interference between European stations. Although the UIR possessed no official character in the 1920s in the sense that states did not bind themselves to adhere to the provisions that were adopted, the necessity of coordinating the use of frequencies to prevent interference resulted in substantial compliance. Later, it became, in effect, the ITU's consultative committee on broadcasting, and its draft plans were adopted at regional conferences as ITU agreements. Eventually, after the Second World War, its functions were assumed by the CCIR and the IFRB.

In 1926 the Technical Commission of the UIR prepared a plan, known as the "Geneva Plan," for the distribution of wavelengths to European stations. The plan alloted eighty-three exclusive and sixteen shared wavelengths in the medium-frequency band to European stations. The chief factors in determining the allotments were:

---

[58] *Minutes of Closing Plenary Session (Ninth Session), November 25, 1927*, International Radio Conference (Washington, 1927), Doc. No. 335, p. 13.

The Product

(1) existing services; (2) area of the country; (3) topography; (4) population density; and (5) economic development. The Geneva Plan went into effect in November, 1927, by agreement among the affected countries.

The 1926 Technical Commission did not undertake the more difficult task of making allotments in the contentious low-frequency band. The next year, however, eleven of the twenty-two countries represented at the UIR Brussels Conference signed an informal agreement establishing such a plan.[59]

A revision of the Geneva Plan was undertaken by the UIR at a meeting at Brussels in January, 1929, necessitated by the many new stations that had been brought into operation. Despite the fact that the plan did not represent a formal agreement among states, substantial compliance existed. Of the 209 stations in operation in 1929, only seventy-two were not adhering to the plan, and forty of those belonged to the U.S.S.R. which was not a member.

At the 1927 Washington conference, the Czechoslovakian delegation suggested the need for a European conference to address the continuing problem of alloting wavelengths to European stations.[60] In accordance with their own suggestion, the Czechs called a conference for that purpose at Prague in 1929. The UIR was invited to assist in conference preparations and twenty-six European administrations were represented.

The procedure adopted in making the allotments was first to make an inventory of the existing usage. Some difficulties were encountered due to the use of non-broadcasting bands by a number of stations—particularly those of the U.S.S.R. The second step was to receive the requests of all the administrations concerning the number and kind of wave-lengths desired. Most of these requests included wave-lengths already in service, as well as the wave-lengths needed for future use, either for stations under construction or stations whose construction was contemplated. Administrations which were more or less satisfied with the allocations made in the Brussels Plan agreed to accept them as the basis of their claims. For

---

[59] See R. Braillard, *The Problems of Broadcasting,* 1930, p. 28.

[60] See International Radiotelegraph Conference, *Documents,* Vol. II p. 533.

the most part, those not satisfied with the plan made special reservations. The task of the Sub-Committee on the Allocation of Wavelengths was to do its best to satisfy these claims.[61]

The Prague conference was the first formal conference among states to effect an allotment plan which produced results acceptable to all the parties. The frequencies were formally reported to the Bern Bureau and the plan was effectively incorporated by reference into the frequency management scheme established under the 1927 International Radiotelegraph Convention. The essential success of the the conference encouraged the use of such an allotment mechanism in the future.

The 1932 Madrid Telegraph and Radiotelegraph Conferences agreed to create the International Telecommunication Union in 1934, and to merge the provisions of the separate conventions into a single telecommunication convention. In effecting the merger, a special chapter for the provisions relating to radio was placed in the Convention, which included a slightly rewritten provision establishing the radio resource right.[62] This right was futher amplified in the General Radio Regulations by a slightly modified version of a similar provision in the 1927 Convention.[63] This provision remained untouched at the next international radio conference held in Cairo in 1938.[64]

It is apparent that the basic international notification and recordation scheme established in 1903 remained essentially unaltered until 1947. At the regional level, however, the allotment planning process continued to be extensively used wherever the use of the

---

[61] John D. Tomlinson, *Supra*, p. 183.

[62] "All stations, regardless of their purpose, must, so far as possible, be established and operated in such a manner as not to interfere with the radio services or communications of either the other contracting governments, or the private operating agencies recognized by these contracting governments and other duly authorized operating agencies which carry on radiocommunication service." *International Telecommunication Convention* (Madrid, 1932), Art. 35.

[63] "Frequencies must be selected in such a way as to avoid, so far as possible, interfering with international services belonging to the contracting countries and operated by existing stations of which the frequencies have already been notified to the Bureau of the Union." General Radio Regulations annexed to the *International Telecommunication Convention* (Madrid, 1932), Art. 7, Sect. 5(1).

[64] See General Radio Regulations (Cairo Revision, 1938) Art. 7, sect. 2 [para. 81], annexed to the *International Telecommunication Convention,* (Madrid, 1938).

radio resource began to get crowded. This work resulted in many of the *a priori* planning principles which were to be later introduced into global ITU forums.

At the 1932 Madrid conference, the wheels were set in motion for the European Broadcasting Conference at Lucerne in 1933, and the Technical Commission of the UIR was called to meet at Brussels in early 1933 to draft a new plan. The crowding among European broadcasting stations had become worse, and the Prague Plan appeared outdated. The Commission began the arduous process of devising an equitable set of principles or formulas for apportioning the available channels. They are worth noting because similar formulas were to be advanced repeatedly during subsequent conferences. Indeed, during the lengthy arguments at the High Frequency Broadcasting Conference at Mexico City in 1950, the participants were reminded that the architect of the U.S.S.R. formula had proposed the same equation sixteen years earlier at Brussels. The UIR recognized six factors which should govern the apportionment process:

(1) The number of existing stations or those under construction or alteration;

(2) The indispensable requirements of a country in which broadcasting was either non-existent, in embryonic development, or in the course of expansion;

(3) Technical and physical considerations;

(4) Political and national considerations: importance of broadcasting as an instrument of government, customs, languages political subdivisions, minorities, etc.;

(5) Conditions already established in a number of countries;

(6) The General Radiocommunication Regulations of Madrid, and the Additional Protocol.[65]

A number of countries submitted other formulas to the 1933 Lucerne conference. None were used explicitly in producing the resulting plan, but rather served as tools of persuasion as to the equity of a particular arrangement during the course of the debates. There was

---

[65] *Documents de la Conference Europeene des Radiocommunications de Lucerne,* (Mai/Juin 1933) Berne: Bureau of the ITU, 1933, pp. 90-91. A report of the entire work of the Brussels Technical Committee can be found at pp. 40 - 161.

significant discontent concerning the resultant Plan, and nineteen formal reservations were entered in the final acts. Nonetheless, the Plan was substantially followed, which demonstrated the utility of the process even under very contentious circumstances.[66]

Although regional broadcasting agreements had not been as necessary outside of Europe, those that existed further established a general consensus that planning conferences were a preferred means of resolving contentious broadcasting allotment disputes. A conference of South American broadcasting organizations was held in Buenos Aires in March, 1934, for the purpose of creating a South American Broadcasting Union (USARD) to perform many of the functions for South American countries that the UIR performed for Europe. The organization was created at Rio de Janeiro in August 1934, and held a conference in March, 1935, for the purpose of establishing a South American plan.

Several years later, the North American countries met at Havana in December, 1937, and established an arrangement which differed significantly from the Lucerne and the South American plans. Part of the agreement left each country free to use any one of the 106 broadcasting channels, insofar as the technical conditions laid down to prevent interference were observed.[67] An additional 106 clear channel stations were apportioned among the countries. Partly because of the greater geographical separation of stations of the countries involved, planning in North America does not appear to have been as difficult a task as elsewhere.

Yet another European planning conference was proposed by the ITU's 1938 Cairo Administrative Radio Conference. In addition, the Technical Commission of the UIR made a proposal that the Cairo conference consider the possibility of a world-wide distribution of frequencies for short-wave, long-distance broadcasting stations, analogous to the Lucerne Plan.[68] This proposal was not acted upon because of a lack of sufficient technical information, resulting in its being deferred to the Atlantic City conferences in 1947.

---

[66] See Tomlinson, p. 204.

[67] Tomlinson, p. 215. The agreement was usually referred to as NARBA (North Atlantic Regional Broadcast Agreement).

[68] See *Cahier des Propositions pour Caire,* p. 76.

Acting on the Cairo recommendation, the European Broadcasting Conference was held at Montreux in 1939, the last major telecommunication conference before the war. A new plan was devised, based on ten basic planning principles. Five countries refused to sign the acts and eighteen entered reservations.

1947 to Present: Contemporary International Rights Vesting Arrangements

On a worldwide level, dissatisfaction with the old notice and recordation process began to grow. It was realized that the mechanism ". . . worked to the disadvantage of those countries which have been hindered, for one reason or another, in the development of their national radio communication networks, and worked to the advantage of those countries which were foresighted enough to make exaggerated claims to frequencies. [I]t is not surprising that after the war, technicians began a search for a more definite and equitable method for the selection of frequencies."[69]

The major focus at the 1947 Atlantic City conferences was on altering the old notice and recordation scheme in several significant ways. First, the existing Berne List (now known as the International Frequency Register), which was the record of all existing rights, would be scrapped. In its place, Committee 6 of the Radio Conference would equitably ". . . engineer. . . a complete and entirely new international frequency list covering all operational assignments in the high-frequency spectrum. [Except,] [i]t is expected that a world-wide assignment plan for high-frequency broadcasting will be prepared by the forthcoming high-frequency broadcasting conference."[70]

Second, new rights would be secured by two means. For high-frequency broadcasting stations, the 1947 Plan would be revised by an international conference which would meet every four years. Subsequent plans would be prepared by a new High Frequency Broadcasting Board.[71] For all other stations, assignments would be

[69] Codding, *The International Telecommunication Union*, p. 192.
[70] U.S.A., Statement. . . with Respect to the Scope and Importance of the Work of International Frequency List Committee, International Radio Conference (Atlantic City, 1947), Doc. No. 78R (May 20, 1947), p. 1.
[71] See U.S.A., Summary of High Frequency Broadcasting Regulations, H.F. Broadcasting Conference (Atlantic City, 1947), Doc. No. 13 Rhf (June 15, 1947), p. 5.

". . . subject to change as a result of findings by the proposed Central Frequency Registration Board."[72]

After a lengthy struggle, this bold endeavor only partially succeeded.[73] It took the next ten years to reach a stable international framework for vesting rights. The scheme adopted was a composite of several different mechanisms, and initially produced substantial dissatisfaction — which was evident from the large number of reservations entered at the adopting conference in 1959.[74] That scheme is largely the one which is used at present.

Contemporary Arrangements

The provision in the Convention which defines the vested right has changed little since it was first drawn up in 1903. Today it reads almost the same as it did in 1932:[75]

All stations, whatever their purpose, must be established and operated in such a manner as not to cause harmful interference to the radio services or communications of other Members or of recognized private operating agencies, or of other duly authorized operating agencies which carry on radio service, and which operate in accordance with the provisions of the Radio Regulations.[76]

The key operative provision in the Radio Regulations is:

Any new assignment or any change of frequency or other basic characteristic of an existing assignment . . . shall be made in such a way as to avoid causing harmful interference to services rendered by stations using frequencies assigned in accordance with the Table of Frequency Allocations in this Chapter and the other provisions of these Regulations, the characteristics of which assignments are recorded in the International Frequency Register.[77]

---

[72] See Ibid., p. 6.
[73] See Chapter Two.
[74] See Additional Protocol to the *Radio Regulations* (Geneva, 1959).
[75] Cf. footnote
[76] *International Telecommunication Convention* (Malaga-Torremolinos, 1973), Art. 35, para. 135.
[77] *Radio Regulations (1982)*, Art. 6, para. 341.

It is important to note that stations *not capable of causing harmful interference to those of another country* need not comply with any ITU arrangements.[78] This is a most important "escape clause" from the ITU's rights' vesting arrangements. It means a nation can do anything it wants with its radio stations so long as harmful interference is not caused to those of another. For that reason, stations operating on frequencies above approximately 50 MHz well within the borders of a country are unlikely to cause interference to stations in another country, and thus will not normally be notified and recorded with the ITU.

Depending on the frequency band and the radio service involved, there are three basic vesting mechanisms which exist today: (1) numerous global and regional plans created and updated at periodic administrative radio conferences; (2) bilateral coordination with the assistance of the International Frequency Registration Board; and (3) unilateral notice and recordation with the IFRB. This composite approach was hammered out during the past thirty years and, with numerous conferences scheduled in the 1980s, is in a state of impending change.

The Plans

The work of the Provisional Frequency Board to produce a plan for the high frequency band resulted in substantial agreement for the maritime and aeronautical mobile services. Thus, the first world-wide *a priori* mechanism was adopted.[79] These plans were amended by several conferences in the 1960s and 1970s, but remain much the same today.[80]

Entirely different concerns gave rise to a planning mechanism for the broadcasting-satellite service. The service has long been a

---

[78] This is a result stemming from another Art. 6 provision: "Members undertake that in assigning frequencies to stations which are capable of causing harmful interference to the services rendered by the stations of another country, such assignments are to be made in accordance with the Table of Frequency Allocations and other provisions of these Regulations." *Radio Regulations* (1982), para. 340.

[79] See *Final Acts of the Extraordinary Administrative Radio Conference (Geneva, 1951)*.

[80] See *Radio Regulations* (1982), Art. 12, Sub-sections IIB and IIC, Art. 16 Appendices 25 - 27.

highly controversial one.[81] In 1977, the broadcasting-satellite service was singled out in the highly desirable 12 GHz band and the radio resource allotted at a conference according to a plan. Specific orbital positions, frequency channels, and geographical areas of coverage were allotted among all nations except those in the western hemisphere.[82] Interim procedures are in effect for the western hemisphere until a 1983 regional administrative conference develops a plan.[83]

Numerous planning mechanisms exist on a regional level. Some are created by ITU administrative radio conferences, others exist within regional organizations and the results notified to the ITU.[84]

---

[81] As early as 1971, the WARC for Space Telecommunications called for a broadcasting-satellite planning conference. See Final Acts of that conference, Res. No. Spa2-2. See also K. Queeney, *Direct Broadcasting Satellites and the United Nations.* The Netherlands: Sitjhoff & Noordhoff, 1978.

[82] See *Final Acts of the World Administrative Radio Conference for the Planning of the Broadcasting-Satellite Service in Frequency Bands 11.7-12.2 GHz (in Regions 2 and 3) and 11.7-12.5 GHz (in Region 1) (Geneva, 1977).* The plan and associated procedures is now contained in Appendix 30, *Radio Regulations* (1982).

[83] See ibid.

[84]The "Regional Agreement for the African Broadcasting Area Concerning the Use of Frequencies by the Broadcasting Service in the Very High Frequency and Ultra High Frequency Bands" came into force in 1964.

The "North-American Regional Broadcasting Agreement" (NARBA) governs the use of broadcasting stations in the band 535 to 1605 kHz, and came into force in 1959.

The "European Regional Convention for the Maritime Mobile Radio Service" adopted the "Copenhagen Plan" for coast stations in the European Maritime Area and came into force in 1950.

The "Regional Agreement Concerning the Establishment of an International VHF Radiotelephone Mobile Service for Rhine Navigation" divides the Rhine area into sections, alloting each a pair of channels. It came into force in 1970 for the five countries involved in the plan.

The "Special Agreement for the Use of Frequencies in the Bands 68-73 MHz and 76-87.5 Mhz by the Broadcasting Service on the one hand and by the Fixed and Mobile Services on the other" was concluded among 20 countries, and sets forth a sound broadcasting and television plan. It came into force in 1961.

The "Regional Agreement for the European Broadcasting Area (Stockholm, 1961)" included extensive plans for sound and television broadcasting in the bands 41-68, 87.5-100, 162-230, and 470-960 MHz. It also included a plan for low-power broadcasting stations in the bands 41-230 MHz. The agreement was concluded among 35 nations and came into force in 1962.

The "Regional Agreement Concerning the Use by the Broadcasting Service of Frequencies in the Medium Frequency Bands in Regions 1 and 3 and in the Low Frequency Bands in Region 1, Geneva, 1975" sets forth an extensive plan, replacing older European and African plans for these bands. It entered into force in 1978.

A half-dozen additional "special arrangements" set forth additional regional plans.

Some important regional planning conferences are scheduled for the 1980s.[85]

Bilateral Coordination

Two important mechanisms exist which can be characterized as bilateral coordination with the assistance of the IFRB. Their origin and nature are quite different.

The work of all the high-frequency broadcasting conferences and special committees which met between 1947 and 1959 was essentially rejected. The mechanism adopted was a flexible means of bilateral coordination with ITU assistance on a seasonal basis.[86] Administrations simply submit their prospective quarterly broadcasting schedules to the IFRB. If conflicts are discovered, the Board notifies the affected parties and assists in resolving the matter. There is, however, a renewed interest in planning high-frequency broadcasting, and a conference to consider the existing vesting arrangements has been called for January 1984.[87]

In 1963 and further in 1971, rights vesting procedures for space radiocommunication were initially fashioned. The use of satellites to relay signals over vast areas of the earth necessitates some kind of cooperative mechanisms to keep them from interfering with other satellites and with terrestrial radio systems. The technology has been changing so rapidly, however, that very flexible procedures seemed appropriate. The method adopted was similar to that devised for high-frequency broadcasting in 1959, i.e., a bilateral coordination process with ITU assistance.[88] The only significant differences in the procedures involve a requirement for advance notification of the satellite system, and limits on the length of time a state can enjoy its rights. The limits, however, are rather vague and indefinite. It is clear they last as long as the coordinated satellite

---

[85]These include: a two-session Region 1 VHF-FM broadcasting conference in 1982 and 1984; a Region 2 broadcasting-satellite conference in 1983; a Region 2 medium-frequency broadcasting conference for the 1605-1705 kHz band in 1986, an African region VHF/UHF broadcasting conference in 1987; and a Region 3 VHF/UHF broadcasting conference in 1988.

[86] See Radio Regulations Art. 10 (1959) [now Art. 17 Radio Regulations (1982).]

[87] See Final Acts of the WARC, Res. 508, (Geneva, 1979).

[88]See *Final Acts of the Extraordinary Administrative Radio Conference to Allocate Frequency Bands for Space Radiocommunications Purposes (Geneva, 1963); Final Acts of the World Administrative Radio Conference for Space Telecommunications (Geneva, 1971).*

system exists, but unclear as to whether the right would continue to vest in a replacement system, particularly if that new system had different characteristics.[89] In addition, provisions exist which affirm that no state enjoys a right in perpetuity for use of the geostationary orbit.[90] However, the practical reality of the procedures could give some *de facto* advantages to the first comer. The later entrant may end up negotiating from an inferior position, seeking such accomodation as other user nations will grant. The burden appears to fall particularly heavily on developing countries.

In view of the rapid changes in technology and the likelihood of space platforms with dynamic radiocommunication systems in the 1990s, the entire problem should disappear as global and regional institutions arise to operate them in lieu of small, less efficient domestic systems. In the interim, however, there is substantial controversy,[91] and a special conference has been called for the mid-1980s to consider new institutional arrangements for general use of the geostationary orbit for radiocommunication.[92]

Notice and Recordation

For all unplanned portions of the radio resource other than high-frequency broadcasting and satellite communication, rights are vested by the old notice and recordation mechanism subject to an initial examination by the International Frequency Registration Board. This result was finally hammered out at the 1959 WARC after twelve years of struggle over the role of the IFRB.[93] The procedure has grown considerably more complex, probably unnecessarily so, [94] with numerous detailed provisions applying to various services and frequency bands, but, in substance, it has remained the

---

[89]See *Radio Regulations,* (1982), Art. 11.

[90]See Final Acts of the WARC. Resolutions 2 and 4 (Geneva, 1979).

[91]See Rutkowski, "Six Ad-Hoc Two: The Third World Speaks Its Mind," *Satellite Communications Vol A, No. 3, (March 1980), p. 22.*

[92] See Final Acts of the WARC, Res. 3, (Geneva, 1979).

[93]See *Radio Regulations,* Art 9, (Geneva, 1959).

[94] The Board has its own internal set of procedures for processing notices. However, Administrative Radio Conferences over the past two decades have apparently been reluctant to allow the Board to provide for administrative details in the internal procedures, electing instead to place them in the Radio Regulations. In addition to making the Radio Regulations far more complex, these details become very unflexible due to the great difficulty in amending them, and often obstruse due to their being hammered out at meetings of more than 100 nations.

same.[95] An administration simply submits a notice to the IFRB of a station which that Administration is authorizing to operate. If the Board finds that the station complies with the Radio Regulations and will not cause harmful interference to any other station previously satisfactorily examined and entered in the Master International Frequency Register, the new station will be entered. The time of entry is determinative of rights against a newcomer. Various provisional, appellate, and insistence procedures exist in the case of unfavorable IFRB findings which may result in eventual entry in the Register with full or secondary rights.

During the past few decades, developing countries have sought assistance from the IFRB in dealing with these procedures.[96] A rather innovative "preferential treatment" procedure was added as a result of WARC 79 negotiations which gives developing countries preferred access to certain frequencies.[97]

Conclusions

For at least the next two decades, frequency management will remain a highly significant function of the ITU. However, the technology is changing exponentially, and it is clear that constant changes will be necessary to provide the necessary international arrangements. The major direction of these changes is suggested by the nature of the technological evolution. Dynamic, digital radio systems under the control of computer processors will take over and revolutionize the radiocommunication environment. Many of these systems will be components of integrated services digital networks which bring together space and terrestrial wire and radio transmission systems under a common network computer intelligence. Most distinctions between radio "services" will fade, as will the technical standards based on those distinctions. Out of this will emerge a common global set of standards appropriate for the new integrated environment.

---

[95]See *Radio Regulations,* Art. 12, (1982).

[96]See, e.g., *Radio Regulations (1959),* paras. 479, 623-34, *Radio Regulations (1982)* paras. 999, 1005, 1184, 1423, 1438-1450, inter alia, *(1982)*

[97] See Final Acts of the WARC, Res. 103 (Geneva, 1979); Radio Regulations, Arts. 12 [paras. 1218, 1260, 1275-1304, 1416], 13 [paras. 1963-66], (1982).

In the rights-vesting area, the next decade or two will be difficult ones to bridge. During this period, individual nations and private corporations are likely to bring great contention to international institutions as they demand their exclusive parcel of the radio resource to operate dedicated hardware for dedicated services. At a point when the technological advancements providing large-scale dynamic facilities can coalesce with new regional and global operating consortia, the individual contentions will become moot. The next two decades promise to be truly exciting ones as the world telecommunications community makes quantum jumps in technological and institutional change.

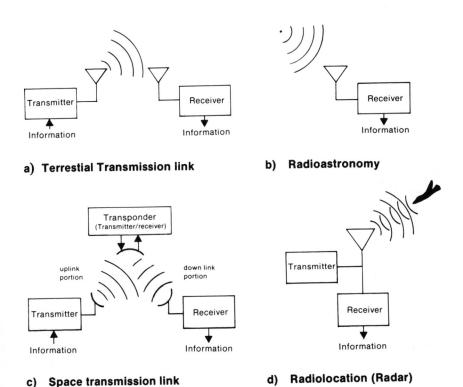

a) **Terrestial Transmission link**

b) **Radioastronomy**

c) **Space transmission link**

d) **Radiolocation (Radar)**

Figure 11.1 Examples of Several Basic Radiocommunication Uses.

TABLE 11.1

A PORTION OF THE TABLE OF FREQUENCY ALLOCATIONS SHOWING ALLOCATIONS OF FREQUENCIES TO SERVICES AND FOOTNOTES*

kHz
19.95 — 90

	Allocation to Services		
	Region 1	Region 2	Region 3
19.95 — 20.05	STANDARD FREQUENCY AND TIME SIGNAL (20 kHz)		
20.05 — 70	FIXED MARITIME MOBILE 448 447 449		
70 — 72	RADIO-NAVIGATION 451	70 — 90 FIXED MARITIME MOBILE 448 MARITIME RADIO NAVIGATION 451 Radiolocation	70 — 72 RADIO-NAVIGATION 451 Fixed Maritime Mobile 448 450
72 — 84	FIXED MARITIME MOBILE 448 RADIO-NAVIGATION 451 447		72 — 84 FIXED MARITIME MOBILE 448 RADIO-NAVIGATION 451 450
84 — 86	RADIO-NAVIGATION 451		84 — 86 RADIO-NAVIGATION 451 Fixed Maritime Mobile 448 450
86 — 90	FIXED MARITIME MOBILE 448 RADIONAVIGATION 447 452		86 — 90 FIXED MARITIME MOBILE 448 RADIO-NAVIGATION 451

kHz
90 — 110

	Allocation to Services	
Region 1	Region 2	Region 3
90 — 110	RADIONAVIGATION 453 Fixed Maritime Mobile 448 454	

ADD 3455A   449   *Additional allocation:* In Bulgaria, Hungary, Poland, the German Democratic Republic, Czechoslovakia and the U.S.S.R., the band 67 — 70 kHz is also allocated to the radionavigation service on a permitted basis.

MOD 3459   450   *Different category of service:* in Bangladesh, Iran and Pakistan, the allocation of the bands 70 — 72 kHz and 84 — 86 kHz to the fixed and maritime mobile service is 165   on a primary basis (see No. 425).

MOD 3456   451   The use of the bands 70 — 90 kHz (70 — 86 kHz in Region 1) and 110 — 130 kHz 162   (112 — 130 kHz in Region 1) by the radionavigation service is limited to continuous wave systems.

MOD 3458   452   In Region 2, the establishment and operation of stations in the maritime radionav-164   igation service in the bands 70 — 90 kHz and 110 — 130 kHz shall be subject to agreement obtained under the procedure set forth in Article 14 with administrations whose services, operating in accordance with the Table, may be affected. However, stations of the fixed, maritime mobile and radiolocation services shall not cause harmful interference to stations in the maritime radionavigation service established under such agreements.

ADD 3461A   453   Administrations which operate stations in the radionavigation service in the band 90 — 110 kHz are urged to coordinate technical and operating characteristics in such a way as to avoid harmful interference to the services provided by these stations.

MOD 3461   454   Only classes A1A or F1B, A2C, A3C, F1C or F3C emissions are authorized for 167   stations of the fixed service in the bands allocated to this service between 90 kHz and 160 kHz (148.5 kHz in Region 1) and for stations of the maritime mobile service in the bands allocated to this service between 110 kHz and 160 kHz (148.5 kHz in Region 1). Exceptionally, class J2B or J7B emissions are also authorized in the bands between 110 kHz and 160 kHz (148.5 kHz in Region 1) for stations of the maritime mobile service.

*From Radio Regulations (1982), Art. 8.

TABLE 11.2

TYPICAL SEGMENT OF THE TABLE OF ALLOCATIONS

GHz
11.7 — 12.75

Allocation to Services		
Region 1	Region 2	Region 3
11.7 — 12.5 FIXED BROADCASTING BROADCASTING- 　SATELLITE Mobile except 　aeronautical mobile	**11.7 — 12.1** FIXED　837 FIXED-SATELLITE 　(space-to-Earth) Mobile except 　aeronautical mobile 836　839　840	**11.7 — 12.2** FIXED MOBILE except 　aeronautical mobile BROADCASTING BROADCASTING- 　SATELLITE 838　840
	12.1 — 12.3 FIXED　837 FIXED-SATELLITE 　(space-to-Earth) MOBILE except 　aeronautical mobile BROADCASTING BROADCASTING- 　SATELLITE 839　840　841 842　843　844	**12.2 — 12.5** FIXED MOBILE except 　aeronautical mobile BROADCASTING
838　840	**12.3 — 12.7** FIXED MOBILE except 　aeronautical mobile BROADCASTING BROADCASTING- 　SATELLITE 839　840　843　844　846	838　840　845
12.5 — 12.75 FIXED-SATELLITE 　(space-to-Earth) 　(Earth-to-space)		**12.5 — 12.75** FIXED FIXED-SATELLITE 　(space-to-Earth) MOBILE except 　aeronautical mobile BROADCASTING- 　SATELLITE　847
840　848　849　850	**12.7 — 12.75** FIXED FIXED-SATELLITE 　(Earth-to-space) MOBILE except 　aeronautical mobile 840	840

(The three digit numbers in the Table are footnote references which further elaborate on the particular allocation.)

chapter 12

DEVELOPMENT ASSISTANCE

The newest service that is performed by the ITU for its members is development assistance. Before the revolutionary change in the international environment brought on by the wholesale gaining of independence by the former colonial territories, the ITU did not concern itself with direct assistance to administrations to help them to improve their telecommunications services. That task was left primarily to the efforts of each member state. Many of the former colonies, however, entered the world scene with very primitive domestic telecommunication systems and international services which were only appendages to the telecommunication services of the former colonial powers, and without the funds necessary to make improvements. As the decisive role that telecommunications can play in the economic, social, and political development of the lesser developed nations has become increasingly obvious, more and more funds have been allotted to the improvement of the telecommunications systems of the newer countries of the world. The ITU has taken on the principal role of channeling these funds into the developing countries, both through the United Nations program and recently, although hesitantly, under its own auspices. The story of this involvement and the problems that it has raised will be the subject of this chapter.

Before proceeding, however, it should be noted that there is an honest difference of opinion over the definition of development assistance. In a way, almost any action that would result in the improvement of the lot of an administration could be classified as development assistance. To a great extent that has been the position of the ITU. The ITU, so the story goes, has been involved in development assistance activities since 1865 when the delegates to the Paris Telegraph Conference shared their experiences, and, in the process, some of the delegates learned how to make improvements in their telegraph services.[1] This catholic view of development assistance is reflected in the use of the term "technical cooperation" by the ITU in preference to "technical assistance" or "development assistance."

Therefore, in order to bring some sort of order to this chapter and to avoid repeating large portions of the other chapters in this and the previous section, we will confine our investigation to assistance of a technical nature which is provided under the auspices of the ITU specifically for the benefit of one sector of its membership, its developing members.

Need and Response

The ITU became officially involved in the United Nations development assistance activities in 1952 when it became a participant in the United Nation's Expanded Program of Technical Assistance. The task of the ITU was to review all projects dealing with "such forms of technical assistance as surveys, expert assistance, fellowships and scholarships, technical conferences, seminars and training centers for the establishment or improvement of telecommunications . . . "[2] The ITU also helped to recruit experts in telecommunications for such projects, to brief them on their missions, and to give advice to the UN on the training of fellows and scholars. One percent of the funds available to the Expanded Program were put aside in 1952 for telecommunication projects, including the administrative expenses incurred by the ITU.[3] The ITU also executed a number of pre-investment surveys for the UN Special Fund after it was created in 1959.

A number of debates were held at the 1959 Geneva Plenipotentiary on the United Nations activities in technical assistance and development assistance in general, resulting in the passage of a number of resolutions aimed at facilitating the ITU's participation in the

---

[1]See, for instance, *From Semaphore to Satellite*, p. 261.
[2]See ITU, *Resolutions of the Administrative Council of the International Telecommunication Union (7th Session, Geneva, April-June 1952)*, Geneva, 1952, p. 134.
[3]Ibid.

United Nations Expanded Program of Technical Assistance and the Special Fund. An addition to Article 4 of the ITU Convention (Purposes of the Union) stated that, in the future, one of the primary aims of the Union would be to "foster the creation, development, and improvement of telecommunication equipment and networks in new or developing countries by every means at its disposal, especially its participation in the appropriate programmes of the United Nations."[4] An addition to Article 13 required that the consultative committees "pay due attention to the study of questions and to the formulation of recommendations directly connected with the establishment, development, and improvement of telecommunication in new or developing countries in both the regional and international fields." The amount of time that was spent on this subject led the U.S. delegate to that conference to observe that "the tremendous interest of the new or developing countries in technical assistance" was "the most significant single trend noted at this Conference."[5]

In reaction to the Geneva decisions and the changes in the development assistance operations with the creation of the Special Fund in 1959, the ITU created a special division of the Secretariat in 1960 to deal exclusively with development assistance matters.

By 1965 the membership of the ITU had changed drastically. Participation in the Montreux Plenipotentiary had increased from 96 to 129 delegations and most of the increase had come from the new developing nations. The new majority was both vocal and insistent, especially in the realm of development assistance.

In the general debates on the subject, the delegates from the developing countries were extremely critical of the level of assistance funds and the administration of those funds by both the ITU and the

[4]ITU, *International Telecommunication Convention, Geneva, 1959,* Geneva, 1960, Art. 4, para. 2. d), Art. 13, para. 1, 3) and Resolutions 24, 25, 26, 27, 28 and 29.

[5]U.S., Department of State, Telecommunications Division, *Report of the United States Delegation to the Plenipotentiary Conference of the International Telecommunication Union, Geneva, Switzerland, October-December, 1959,* TD Serial No. 905, Washington, D.C., January 15, 1960 (mimeo), pp. 5 and 16. For this American delegate, the trend was not all that bad. After summarizing all of the changes in the Convention and all of the resolutions and recommendations that had been passed, the U.S. Delegate concluded with: "It is quite obvious from the Conference that hereafter the question of technical assistance will be one of the most important activities of the Union and that, in this respect, it will join the ranks of the other Specialized Agencies of the United Nations, which from the first have given technical assistance the priority that it deserves." Ibid., p.16.

UNDP. Three radical proposals were made by the new majority to overcome what they saw as the program's major deficiencies. First, an ITU development assistance fund should be created, separate from that of the United Nations. Second, a number of regional ITU offices should be established in less developed areas of the world to help local administrations with their telecommunication problems. And third, a new technical assistance organ of the ITU should be created which would be similar to the consultative committees in that it would be independent of the Secretary-General and have its own elected director.

The developing countries were able to push through a number of resolutions aimed at increasing ITU activity in development assistance and making that participation more effective. These resolutions included a decision to recruit four telecommunication specialist engineers for the General Secretariat to help the developing countries "on difficult, specific national problems in telecommunication development as they arise, in particular in the fields of network planning, preparation of specifications, and evaluation of systems," directing the Secretariat to investigate training methods for the personnel of these countries, and to hold seminars for them. All the permanent organs of the Union were to help in the dissemination of scientific knowledge, and other useful advice to the developing countries through various publications.[6] The new nations failed to gain acceptance of their three major proposals, however. All were defeated in committee, although the proposal to create an independent ITU development fund lost by only one vote. The activities of the new majority at Montreux led the U.S. delegate to observe: "It may be stated that if there was one dominant note at the Conference, it was the oft-stated belief that the ITU has as one of its basic obligations the requirement to assist the new and developing nations with their individual telecommunication problems," an activity the delegate believed was a decided shift away from the purpose of previous conferences which was "to expand and improve the global telecommunication networks maintained by the leading telecommunication administrations of the world. . . ."[7]

---

[6]The quotation is from ITU, *International Telecommunication Convention, Montreux, 1965,* Geneva, circa 1966, Resolution No. 29. See also Resolutions Nos. 27, 28,

[7]U.S., Department of State, Office of Telecommunications, *Report of the United States Delegation to the Plenipotentiary Conference of the International Telecommunication Union, Montreux, Switzerland, September 14 to November 12, 1965,* TD Serial No. 973, Washington, D.C., December 15, 1965 (mimeo), pp. 8 and 9.

The ITU became a partner in the United Nations Development Program (UNDP) when it absorbed the Special Fund and the Expanded Program of Technical Assistance in 1965/66. Under the UNDP, the ITU and the other executing agencies work to help local governments to plan the telecommunication sector of their Country Program to be submitted to the UNDP. After a project has been assigned, the ITU is mainly in charge of recruiting experts, placing fellows in training programs, and purchasing any necessary equipment.

The Malaga-Torremolinos Plenipotentiary of 1973 was almost completely dominated by the concerns of the developing countries, with special emphasis on development assistance. Some thirteen major proposals dealing with this subject were discussed in committee and plenary meeting, eleven of which were approved. Many of these were extensions or updates of recommendations passed in Montreux, but there were some innovations. Resolution 19, for instance, requested that the Union make a special effort to help meet the needs of the "least" developed countries and Opinion 2 asked the developed countries to "take into account the requests for favorable treatment made by developing countries in service, commercial, or other relations in communications."[8] The most controversial, however, was the decision to create a special technical cooperation fund 'to meet the needs of the developing countries who submit urgent requests for assistance to the Union."[9] The Administrative Council was instructed to supervise the management of the fund and to take all steps necessary to ensure its efficient operation and growth. There was some sentiment in favor of financing the new fund from the regular budget but this was defeated decisively in favor of voluntary contributions. A number of countries argued against the whole concept of the fund, suggesting that all technical assistance should be coordinated through the UNDP, where telecommunication developments would be weighed along with the other needs of the developing countries. The United States was one of the dissenting countries, and it vowed that if the fund was created it would not contribute to it.

The whole subject of the creation of ITU regional offices in the lesser developed areas of the world was raised again at Malaga-Torremolinos, as was the issue of a new International Consultative

[8]Ibid., Resolution No. 19 and Opinion No. 2.
[9]Ibid., Resolution No. 21.

Committee for Technical Cooperation. The regional office concept was defeated in committee by a vote of 36 for, 36 against, and 4 abstentions, but when put to a secret vote, won by 47 to 42 with one abstention. When brought to the floor in the plenary assembly, however, it was defeated again in a secret vote of 45 in favor, 68 opposed, and four abstentions. The proposal to create a new consultative committee for development assistance problems received so little support in committee that it was not brought to the floor in the plenary assembly.[10]

The debate as to the exact nature of the participation of the ITU in the development assistance process is far from complete. A large number of members of the Union feel that the present situation is not satisfactory, and many crucial issues have not been dealt with in any great depth. The close vote on the problem of regional offices alone indicates that development assistance will be an important subject at the 1982 Plenipotentiary. In addition, in 1979 the Administrative Council set up a working group on the future of ITU technical cooperation activities which should have some interesting information to present to the 1982 Plenipotentiary.

Objectivies of the Program

The ITU divides its development assistance objectives into three categories: the promotion of the development of regional telecommunication networks, the strengthening of telecommunications technical services and the development of human resources.

There are four regional telecommunication networks in progress: the Pan-African Telecommunication Network (PANAFTEL) and networks for the Americas, Asia, the Middle East, and the Pacific. The best known is PANAFTEL, and the ITU's involvement in that network has been extensive. The concept of creating a coordinated African telecommunications network, which had been raised by a numbɜr of African leaders, had its beginning at a meeting of the ITU's Regional Plan Committee for Africa held in Dakar in 1962. While telecommunication experts were responsive, it took some time to convince other responsible government authorities. The breakthrough came in 1965 when the ITU convinced the United Nations

[10]See U.S., Department of State, Office of Telecommunications, *Report of the United States to the Plenipotentiary Conference of the International Telecommunication Union, Malaga-Torremolinos, Spain, September 14 - October 25, 1973,* TD Serial No. 43, Washington, D.C., circa 1973 (mimeo), pp. 40-41 and 44-45.

Economic Commission for Africa (ECA) that such a network would be beneficial to the overall development of the region.

At the 1967 ECA Conference in Lagos, a joint ITU/ECA expert mission was established to make "a thorough and systematic survey of the technical and financial requirements" of such a network and "to establish in detail the precise requirements of an integrated, high quality, and economically justifiable network which, in conjunction with the existing development plans of the administrations, would meet the basic needs of the African continent."[11] Funds were obtained from the UNDP to finance the joint ITU/ECA mission, which begain its studies in 1968 in Eastern and Northern Africa and in 1969 in Central and Western Africa. By 1972, an ITU official could report that these studies "covered well over 80% of both the area and population of Africa and identified, on the basis of the traffic forecasts derived from studies of population distribution and growth of foreign trade and national income, the routes which would constitute the most economical plan for the Pan-African telecommunication network, including both terrestial and satellite links."[12]

A coordinating committee was created in 1972 consisting of the ITU, ECA, and the African Development Bank (ADB), which began the search for funds. A highlight in this search occurred in 1974 in Geneva when representatives of a number of multinational and bilateral financial organizations met at the suggestion of the committee to discuss how to finance the network, which was soon forthcoming. As stated by the ITU's Secretary-General, "it is notable that . . . 75% of the required investment of 150 million U.S. dollars (1974 values) was secured within one year. This success was largely due to the detailed studies, made by the consultants working under ITU supervision, of potential traffic and profitability of the proposed network."[13]

Since construction of PANAFTEL was begun, the ITU continues to be a major partner in the Coordinating Committee, which now consists of the three original members, the Organization of African Unity (OAU), and the Pan-African Telecommunication Union

---

[11]From ITU, *Pan-African Telecommunication Network,* Geneva, 1979, p. 7.

[12]Ibid.

[13]Mohamed Mili, "Importance of Telecommunications for Developing Countries," *The Indian and Eastern Engineer,* 119th Anniversary Edition, as quoted in ITU, *Teleclippings,* No. 525, 29 May 1978, p. 5.

(PATU). In addition to its supervisory duties involved in that partnership, the ITU has been active through the UNDP in providing experts, helping to train local personnel, and purchasing equipment."[14]

In 1980, the ITU was responsible for seventy-nine expert missions that were carried out in relation to the promotion of regional telecommunication networks and their integration into a worldwide system.[15]

The second major endeavor, strengthening of telecommunication technical services, involves helping administrations to solve planning, technical, and administrative problems involving almost all branches of telecommunication. It includes the recruitment of experts, the placing of fellows, and the purchase of equipment. While the experts—there were 198 expert missions in 1979—are primarily advisers, in some cases they participate actively in the execution of projects where the local government lacks the necessary personnel. The following examples for Africa in 1980 portrays the scope of this activity:

- Establishment and operation of a Network Rehabilitation Unit, Cameroon;
- Installation of rural radio-call services, Ethiopia;
- Assistance in the rehabilitation of networks, Equatorial Guinea;
- Assistance in development planning, Guinea;
- Development of radio monitoring and frequency management facilities, Malawi
- Advisory services in planning and establishing a centralized maintenance system for Casablanca, Morocco;
- OPAS services in telecommunication planning, Mozambique;
- Assistance in an integrated development program (telecommunication, planning, rural communications, training), Rwanda;

---

[14]For a more detailed account of PANAFTEL and the ITU's involvement see Sajo Magan-Nyameh Camara, *the Pan-African Telecommunications Network.* Thesis for the Master of Science in Telecommunication, University of Colorado, 1981.

[15]See ITU, Administrative Council, 36th Session, Geneva, June 1981, Addendum No. 3 to Document No. 5621-E, 29 April 1981, p. 6/13. Other figures in this section on technical cooperation activities in 1980 are from this same document unless otherwise noted.

- Provision of OPAS engineers to the P & T Administration, Swaziland
- Assistance in the restoration of rural radio-call services and in the rehabilitation of the telecommunication network, Uganda; and
- Assistance in development planning, Zaire.[16]

The third area of concern, and the largest from the point of view of expenditure for field operaions, is the development of human resources. This type of assistance involves training of staff in the developing nations to meet various manpower needs. In 1980 this type of assistance "consisted in the establishment and/or improvement of national or multinational training institutions, as well as in-service and on-the-job training, the organization of short-term specialist meetings and seminars, and the implementation of fellowships."[17] The ITU is especially proud of its efforts to help establish a high-level telecommunicaiton training school (*Ecole Superieure Multinationale des Telecommunications*) in Dakar, Senegal, which will train individuals from the French-speaking nations of Africa, and the African Regional Advanced Level Training Institution in Nairobi, Kenya, to provide for the training needs of Eastern and Southern African countries.

In 1980, efforts in these three fields included the arrangement of 630 expert missions, the sending of 650 fellows abroad for training, and the purchase of some $8,903,467 worth of equipment. Total expenditure by region in 1980 was as follows:[18]

The total expenditures of the ITU on development assistance projects for the fifteen year period 1966-1980 is found in Table 12.1.

Financial Resources

The bulk of the funds for the technical cooperation projects discussed in the preceding section of this chapter come from the UNDP. The UNDP pays for the projects, and, as mentioned earlier, pays the ITU a fee for administrative costs that are involved in its participation; these are used mainly to support the Technical Cooperation

---

16Ibid., p. 6/9.
17Ibid., p. 6/11.
18Ibid., p. 6/14.

Division of the General Secretariat. The total amount allocated by the UNDP in 1980 for telecommunication projects to be administered by the ITU was $27,539,536, or 82.6 percent of all the development assistance funds available to the ITU.

As is the case with other international organizations, the ITU also receives financial assistance from the Funds in Trust and the Associate Expert Scheme. The Funds in Trust arrangement is one in which member countries and other entities can "hire" the ITU, through its Technical Cooperation Division, to aid them or other member countries of their choice in technical cooperation projects. In 1980, for instance, there were Funds in Trust arrangements between the ITU and the following countries to provide assistance: Guatemala, Honduras, Kuwait, Netherlands Antilles, Niger, Portugal, Saudi Arabia, Surinam, and the Yemen Arab Republic (P.D.R.). This assistance took the form of thirty-seven expert missions, twenty-four fellowships, and the equipment valued at $1,858,729. In addition, the Netherlands financed a project in Surinam, Germany (F.R.)financed the beginning of a "prefeasiblity study" of the application of modern technology for integrated rural development, and Sweden financed a project manager for a plan to draft teletraffic engineering courses for distribution to the developing countries. Further, in 1980, the ITU had an arrangement with the European Economic Community to find the assignment of two training experts to the National Training Center in Botswana and Swaziland, with the Arab bank for Economic Development in Africa, for six experts to conduct a feasibility study on African telecommunications, with the Economic Community of West African States (ECOWAS) to fund twelve expert missions on the socioeconomic development of the telecommunications sector of West Africa, and with Gulf Vision (an agency involving the governments of Saudi Arabia, Bahrain, United Arab Emirates, Iraq, Kuwait, Oman and Qatar) for an expert to make a radio propagation survey project in the Gulf States. The total amount that the ITU was able to use for technical assistance in 1980 from this source was $4,935,329 or 14.8 per cent of the total.

The Associate Expert Scheme is one in which the more advanced governments provide, at their own expense, young engineers to work with experts recruited by the ITU on various development projects. For the most part, the Associate Experts are used in the training of telecommunication personnel. In 1980, twenty-five Associate Expert missions were provided by Denmark, Germany (F.R.), Netherlands, Sweden, and Switzerland at a cost of $864,287 (2.6%). As noted earlier, the ITU also receives a fee in both of these arrangements to cover administrative costs.

Other sources of income for telecommunication development projects have always been minor, including income from the ITU's own development fund. As reported earlier, the contributions to this fund have never been large; the United States still refuses to contribute to it.

Expenditures from the ITU Development Fund in recent years amounted to:

In general, the bulk of contributions to the fund have come from the developed countries of Western Europe, Scandinavia, and Japan. Very small amounts come from other sources. In 1980, for instance, the ITU reported the receipt of 25,230 Swiss francs from the Netherlands, and an expenditure of 23,277 Swiss francs to finance two short term fellowships for nationals of Comores and Tanzania and equipment for the Congo. The entire expenditure from the ITU Development Fund in 1980 amounted to less than one tenth of one per cent of the total amount devoted to technical assistance.

The Role of the ITU

In view of the small amount of development funds at the disposal of the ITU, and the number of countries which need help in order to create an efficient national telecommunication network, it is obvious that it would be impossible for the ITU to change the overall situation to any degree by a direct application of those funds. The question then arises, just what is the role of the ITU in development assistance?

In the first place, the ITU can help countries determine the extent of their telecommunication needs and create a plan of action to meet those needs. To a great extent, this is what the Joint Plan Committees of the CCIR and CCITT are doing for the various regions of the world such as Africa. The ITU has also made a number of feasibility studies, at the request of individual countries.

Once the country or countries in question have made the improvement of telecommunications a priority item in their overall national objectives, the ITU can help them in making important pre-investment studies. On the basis of these studies the developing country, with the help of the ITU, can turn to international funding agencies and bilateral funding sources to obtain the funds necessary to carry out their plans.

When the actual projects have been funded and are underway, the ITU has a supervisory role to play, including re-evaluation of the overall plan in the light of progress that is achieved. This activity could be carried out by the ITU alone, but is normally carried out by the ITU, the funding agency, and other agencies that the ITU has encouraged to participate. The ITU can also provide expert advice on any specific aspect of such projects, can trouble-shoot when requested, and can help in the preparation of tender specifications.

The ITU also has a continuing role in developing the personnel needed to operate the telecommunication facilities created through the administration of fellowships, the creation of training facilities, and the purchase of training equipment. To this can be added such activities as the production of the GAS manuals by the consultative committees, seminars run by the IFRB on frequency management and the use of frequency spectrum and the geostationary orbit, and other seminars on telecommunications that are sponsored by the ITU.

It should be noted, however, that the bulk of the funds for these activities comes from the UNDP. Thus, in order to be financed, the activity in question must be a part of a country UNDP program. The country and the UNDP, therefore, decide how much of the overall UNDP funds will be expended on telecommunications. The ITU can encourage developing countries to make the improvements of their telecommunications an important part of the overall UNDP plan, and does what it can to bring the importance of an efficient tele-communications network to developing countries. However, in the final analysis, it is the country in question and the UNDP which decide how much emphasis to place on telecommunications.

The amount of freedom of action left to the ITU in projects sponsored by the Funds-in-Trust and Associate Experts is also extremely limited. About the only area where the ITU is free to determine where development funds should be spent is in the case of the special ITU development fund. However, the amount made available to this fund is so small that it is not really important in our overall calculations.

The role of the ITU is, therefore, primarily one of administrator of funds of other agencies. While the ITU does, and should, try to encourage countries to make telecommunications a priority in overall development plans, and it will help countries to plan the telecommunication sector that has been included in these plans, beyond that its role is primarily that of adviser, counselor, and friend.

Conclusions

At the insistence of its developing country majority, the ITU has made development assistance a priority function. There are a number of sectors where the ITU acting alone can influence the use to which this assistance is given. However, its development assistance function is primarily as administrator of UNDP funds. The final decision on the amount that a country receives from this fund and the areas where it will be used depends on the UNDP and the country in question. The desire for a strong ITU input into the original country's plans for development is one of the reasons that some ITU officials have promoted the concept of a large number of ITU regional offices. The other reason, of course, is that developing countries want the expert advice that would be made available.

The proposals for the creation of an ITU development fund have several bases. In the first place, an independent ITU fund would give the ITU a freer hand in determining what projects should be supported. It also would give more funds to telecommunications, allow for quicker reaction to needs than is available from the UNDP, and free telecommunication development from fluctutations in the UNDP funds such as those which occurred in the latter part of the 1970s. For the Secretariat, it would mean a bigger staff.

Changes in the system are opposed largely by those developed countries, who feel that the amount being contributed to development assistance is adequate and that it should all be coordinated by an agency such as the UNDP, which can weigh the various needs of a country or region and put telecommunications in its proper perspective.

Because development assistance is such a high priority item in the minds of the people who run the governments in the developing country majority, the amount of telecommunications development assistance, the uses to which it should be put, and the role of the ITU will continue to be a major concern in ITU conferences and meetings.

TABLE 12.1
DOLLAR VALUE OF ALL ASSISTANCE PROVIDED TO DEVELOPING COUNTRIES BY REGION IN THE PERIOD 1965 TO 1980 (IN THOUSANDS)*

Year	Africa	Americas	Asia & Pacific	Europe & Middle East**	Inter-Regional	TOTAL
1965	1,185	562	1,570	35	7	3,358
1966	1,601	815	1,913	66	72	4,467
1967	1,762	878	2,815	78	43	4,945
1968	2,072	933	1,775	83	19	4,883
1969	2,175	950	1,864	166	70	5,226
1970	1,865	1,442	2,501	244	0	6,051
1971	2,751	1,749	2,980	217	1	7,697
1972	3,127	1,879	3,625	367	50	9,048
1973	3,861	2,204	4,401	276	***	10,743
1974	4,312	2,785	5,279	238	75	12,689
1975	6,093	4,517	4,786	3,295	147	18,383
1976	7,196	3,778	4,065	4,785	418	20,242
1977	5,472	2,685	3,460	4,976	503	17,096
1978	5,910	2,801	5,661	6,797	446	21,615
1979	7,723	5,024	5,831	7,174	212	26,064
1980	11,271	4,782	7,829	9,156	306	33,353

*From Telecommunication Journal, Vol. 40, No. VIII (August, 1973), p.445-446; Report on the Activities of the International Telecommunications Union, 1973, p. 48; 1974, p. 66; 1975, p. 57; 1976, p. 50; 1977, p. 63; 1978, p. 58; 1979, pp. 70-71; and ITU. Administrative Council, 36th Session, Geneva, June, 1981, Addendum No. 3 to Document No. 5621-E, 29 April, 1981, p. 6/14.
**Until 1974 listed only as Europe.
***No figures given.

chapter 13

THE INFORMATION FUNCTION

As the major international mechanism for cooperation in the field of telecommunications, the ITU provides a highly valuable information service for its members. The Union collects, processes, and disseminates massive amounts of information concerning the administration and operation of telecommunications throughout the world. Such information is used to guide national regulatory policy and to assist in the administration of domestic and international telecommunication systems.

The ITU's information function is carried out through a wide variety of activities. These include "bulletin board" excercises, such as the publication of various Secretariat circulars and the *Journal*, the operation of the library and archives, the production of lists of call signs for ships and the annual handbook of telecommunication statistics, and the maintenance of the International Frequency List.

The information exchange also takes place informally when delegates talk to each other during conferences and meetings in the convention hall, when delegates meet after hours, or during lunch, dinner, or coffee breaks. Delegates learn more from each other than appears in the formal documents. Many of the officials of the ITU give formal and informal talks to members of national administrations and are often participants in meetings of representatives of industry and international organizations of various kinds with an interest in telecommunications. Further, the headquarters of the ITU in Geneva receives a steady flow of visits from officials of telegraph, telephone, and radio administrations as well as private operating agencies and other international organizations with an interest in telecommunications matters. The exact amount and the exact nature of this flow of information is impossible to measure with any degree of accuracy. From all indications, however, it is far from negligible.

This chapter will focus on three aspects of this information function which have not been described in other chapters in any detail: the publications produced by the Secretariat, the TELECOM exhibitions, and an important new development, the creation of an ITU documentation center.

Publications

The compilation and distribution of ITU publications, in the words of the ITU's Deputy Secretary-General, "guarantees the continued flow of international communications on a world-wide basis."[1] These publications are of two major types: those available to the public for a price, and those normally available only to the delegates to conferences and meetings. We will begin with the latter.

Information Associated with Conferences and Meetings

A conference of the ITU is much more than discussions between delegates, debates, and voting. Proposals must be circulated in advance of the meeting in order for administrations to prepare their stand on issues. In order to save the valuable time of the delegates, the proposals must be correlated according to the issues involved. During the conference itself, directives from those in authority (chairmen and vice chairmen) must be distributed, and delegates must be informed of what went on in all committees and plenary meetings, and the final product must be put into a usable form. In the process, it is necessary to make the words that are spoken and written understandable to as many as possible. In addition to the contribution of the very visible interpreters in their booths above the delegates' heads, there are the hours of effort from the legion of hard-working, expert translators in offices in the Secretariat building. All of this is mainly the reponsibility of the Department of Conferences and Common Services of the General Secretariat.

---

[1] From a speech by Richard A. Butler entitled "World Telecommunication Development and the Role of the International Telecommunication Union," delivered to the International Conference on Trans-National Data Regulation in Brussels, 9 February 1978.

Most of the documentation that is thus provided, it must be admitted, is determined by agencies other than the General Secretariat, but it is the Secretary-General who determines its form, and thus its emphasis, and he can and does often provide information on his own that he feels will aid the meetings in making decisions. In any case, the focal point of a delegate's day is the morning trip to his mail box to pick up the documents that will inform him about what will be discussed and on which he must often rely when forced to make a decision that could profoundly affect his government. In the end, these documents also inform the member administration of what has been accomplished during the conference and what it must do to adapt its system to the changes approved by the delegates.

As an example of the work involved, for WARC 79 in 1979 the Language Division of the department translated 12,401 pages of text, the Typing Pool produced 46,345 pages, and the Publications and Reprography Division made a total of 49,420,253 offset machine runs. In all, the Conference Documents Service distributed some 4,367,380 documents during the course of WARC 79.[2]

Similar informational support services are provided for meetings of the Administrative Council, the IFRB, and the plenary assemblies and working groups of the CCIR and the CCITT. In 1979 the Language Division of the Conferences and Common Services Department translated 2,209 pages for the Administrative Council, 1,910 for the IFRB, 911 for the CCIR, and 20,778 for the CCITT. The Typing Pool produced 5,872 pages for the Council, 1,115 for the IFRB, 754 for the CCIR, and 36,280 for the CCITT. Offset machine runs for the Council in 1979 were 1,505,509; 1,176,311 for the IFRB, 562,240 for the CCIR, and 25,354,135 for the CCITT.[3]

Additionally, there are the officials from the Secretariat who act as secretaries to the various committees and subcommittees of conferences and meetings, bringing with them their knowledge of procedures and of the working of the ITU. The Secretary-General himself acted as the Secretary of WARC 79 and was assisted by A.S. Winter-Jensen, head of the Conference Preparation Division of the De-

---

[2]ITU, *Report on the Activities of the International Telecommunication Union in 1979*, Geneva, 1980, pp. 13-14.
[3]Ibid.

partment of External Relations, G.C. Brooks and A.A. Matthy, the senior members of the specialized secretariat of the IFRB, and the Union's Legal Councillor. H. Pouliquen, Technical Adviser to the Head of the Department of External Relations, served as secretary to the plenary assemblies and the Heads of Delegations meetings (Committee 1); senior officials from the same department were the secretaries for Committee 7 (General Administrative Committee), Committee 8 (Restructure of the Radio Regulations Committee), and Committee 9 (Editorial Committee). The head of the Conference Preparation Division of the Department of External Affairs also served as the secretary to Committee 2 (Credentials Committee); the Chief of the Finance Department, R. Prelaz, was the secretary to the Budget Control Committee (Committee 3); and senior officials from the technical section of the specialized IFRB secretariat acted in the same position for the Technical Regulations, Frequency Allocations and Regulatory Procedures Committees (4, 5, and 6). These individuals were constantly at the ready to advise and inform the delegates involved.[4]

The delegates to WARC 79 were also able to call on the experience and expert knowledge of the numerous members of the Secretariat who were detailed to the conference for other purposes. The remainder of the Secretariat was available, if needed, just across the street from the conference center. This is an important consideration when the event is held in Geneva, as are most meetings.

Once the conference or meeting has been completed, the Secretariat prepares the final documents, including translation into the official languages, and distributes them to the member administrations. For the plenipotentiary conferences the two major such publications are the *International Telecommunication Convention,* which includes the texts of all the resolutions, recommendations and opinions that were passed, and the minutes of the plenary meetings. The Convention drafted by the 1973 Malaga-Torremolinos Plenipotentiary amounted to 269 pages, including index, and the minutes of the plenary sessions totaled 503 pages in the English texts.[5]

---

[4]See ITU, *World Administrative Radio Conference, Geneva, 1979,* Document No. 982-F/E/S (1 February 1980), p. 96.

[5]See ITU, *International Telecommunication Convention, Malaga-Torremolinos, 1973,* Geneva, 1974, and *Minutes of the Plenipotentiary Conference of the International Telecommunication Union, Malaga-Torremolinos, 1973,* Geneva, 1974.

For the world administrative radio conference, the Secretariat publishes, revises, and markets the Radio Regulations, which filled two volumes and over 700 pages in the 1976 edition of the 1959 Regulations, and the Minutes of the Plenary Meeting, 389 pages in 1979. The Secretariat does a similar job for restricted radio conferences, including bringing the Radio Regulations up to date as a result of the decisions that are made.[6]

A similar service is performed for the administrative telegraph and telephone conference, although, as a result of the decision to let the CCITT establish technical and operating norms, the Telegraph and Telephone Regulations are only published as part of the adopting conference.[7]

The Secretariat publishes the resolutions and decisions that the Administrative Council passes at its annual meeting dealing with the operation of the ITU and its organs, distributes it to the member countries, and puts the remainder on sale for those who might be interested.

To know the latest technical and operational standards which will enable an administration to interconnect with the international system and often to utilize its own telecommunication network in the most efficient manner, it is necessary to have the most current edition of the documents of the plenary assembly of the CCIR and the CCITT. These are on sale in three different languages at the ITU headquarters in Geneva. The CCITT CAS manuals can also be purchased.

The ITU also publishes a biennial edition of the IFRB's International Frequency List, which is essential in frequency use planning for most telecommunication administrations, and a weekly IFRB circular, which will permit an administration to keep up with the IFRB's work in international frequency management.

The General Secretariat is also responsible for a number of publications which keep member administrations aware of developments

---

[6] See ITU, *Radio Regulations, Edition of 1982* Geneva, 1982 and *Minutes of the Plenary Meetings of the Administrative Radio Conference of the International Telecommunication Union, Geneva, 1959,* Geneva, 1960.
[7] The 1973 Telegraph Regulations and the Telephone Regulations and all the Resolutions, Recommendations, and Opinions amounted to a grand total of only 53 pages.

in international telecommunications and within the ITU itself. Among these are the Secretariat's *Notifications,* twelve of which were produced in 1979, *General Secretariat Circulars,* forty-four in 1979, and periodic *Lists of Publications.* The Secretariat also publishes an annual *Financial Operating Report* and an annual *Report on the Activities of the Union.*

Probably the best known publication to come out of the ITU headquarters is the monthly *Telecommunication Journal,* also published in three languages. This journal contains information concerning the various organs of the Union and articles on telecommunications from members of the Secretariat, individuals from member administrations, and sometimes, from private telecommunication experts. The wide audience of this journal results in extensive and elaborate advertisements placed in it by manufacturers of telecommunication equipment and suppliers of telecommunication services.

Service Documents

The General Secretariat is also responsible for the compilation, publishing, and distribution of a number of service documents which are extremely important to administrations in the day-to-day conduct of telecommunication affairs, and without which the operation of many of the world's telecommunication services would be difficult, indeed, if not impossible. The 1973 Telegraph and Telephone Regulations, for instance, require the Secretary-General to publish the following service documents:[8]

1. *Yearbook of Common Carrier Telecommunication Statistics*
2. *Transferred Account Booklet*
3. *International Credit Card for Telegraph Services*
4. *Codes and Abbreviations for the Use of the International Telecommunication Services*
5. *List of Destination Indicators for the Telegram Retransmission System and of Telex Network Identification Codes*
6. *List of Telegraph Offices Open for International Service*
7. *List of Cables forming the World Submarine Network*

---

[8] ITU, *Final Acts of the World Administrative Telegraph and Telephone Conference Geneva, 1973, Telegraph Regulations, Telephone Regulations,* Geneva, 1973, pp. 92-94.

8. *List of Point-to-Point Radio Telegraph Channels*
9. *List of Definitions of Essential Telecommunication Terms*
10. *Telecommunication Statistics*
11. *Routing Table for Offices Connected to the Gentex Service*
12. *Transferred Account Table*
13. *Table of International Telex Relations and Traffic*
14. *Table of Service Restrictions*
15. *Tables of Telegraph Rates*
16. *List of International Telephone Routes*

The 1979 Radio Regulations require the Secretary-General to publish the following service documents: [9]

1. *The International Frequency List*
2. *List of Fixed Stations Operating International Circuits*
3. *List of Broadcasting Stations Operating in Bands below 26,100 kHz*
4. *List of Coast Stations*
5. *List of Ship Stations*
6. *List of Radiodetermination and Special Service Stations*
7. *Alphabetical List of Call Signs and/or Numerical Table of Identities of Stations used by the Maritime Mobile Service and Maritime Mobile-Satellite Service (Coast, Coast Earth, Ship, Ship Earth, Radiodetermination and Special Service Stations), Ship and Ship Earth Stations Maritime Mobile Service Identities and Selective Call Numbers or Signals, and Coast and Coast Earth Stations Maritime Mobile Service Identities and Identification Numbers or Signals*
8. *Alphabetical List of Call Signs of Stations other than Amateur Stations, Experimental Stations and Stations of the Maritime Mobile Service*
9. *List of International Monitoring Stations*
10. *List of Stations in the Space Radiocommunication Services and in the Radio Astronomy Service*
11. *Map of Coast Stations which are open to Public Correspondence or which Participate in the Port Operations Service*
12. *Chart in Colours showing Frequency Allocations*
13. *Manual for Use by the Maritime Mobile and Maritime Mobile-Satellite Services.*

[9]ITU, *Final Acts of the World Administrative Radio Conference, Geneva, 1979*, pp. RRN24-1 to RRN24-6.

The contribution of publications to the information function of institutions depends to a great extent on their availability to those who need them; for many, availability depends directly on cost. The rising cost of ITU publications is a continuing preoccupation of ITU authorities.

The 1952 Buenos Aires Plenipotentiary introduced the principle that the expenses of ITU publications including the staff costs involved, should be met from the proceeds of the sale of documents, (the 1947 Convention had not included any provision dealing with the cost of ITU publications.) As the cost of materials and labor continued to rise, the ITU was forced to increase the price of ITU documents at fairly frequent intervals. A revaluation of the Swiss franc in the early 1970s resulted in a price increase that forced ITU authorities to look for methods of reducing, or at least holding down, the price of ITU publications. The sale of documents had fallen drastically; some administrations were buying fewer copies and reproducing them themselves, and other administrations were simply trying to do without.[10]

At its 31st meeting the Administrative Council decided on a number of measures to keep the price within the range of all member administrations. It agreed to advertise its publications more extensively and to see if economies could be made by using outside printing facilities. Perhaps the most important measure was the transfer of nine staff positions from the publications budget to the ITU's ordinary budget. The Administrative Council had frowned on such a move in the past because many ITU documents are sold to private operating agencies, and thus such a move would tend to subsidize their purchases. The Council, however, came to the conclusion that such a move was necessary, although they did cut the number of posts to be transferred from the seventeen proposed by the Secretary-General to nine.[11]

The Administrative Council has continued to transfer posts from the publications budget to the ordinary budget (four more in 1979) and to explore methods of reducing costs even further, including

---

[10]See the account in U.S., Department of State, Office of International Communications Policy, *Report of the United States Representative on the Administrative Council of the International Telecommunication Union, 31st Session, Geneva, Switzerland, June 14-July 2, 1976*, TD Serial No. 72, Washington, D.C., 1976 (mimeo), p. 5.
[11]Ibid.

better marketing and even the possibility of reducing the number of languages in which they are published. A special effort is being made to reduce the cost of the final reports of the CCI plenary assemblies.

In the process, the Secretary-General and the Administrative Council have also been investigating the possibility of publishing documents in other than the printed form. The first major step in this direction was the decision of the 1981 Administrative Council to begin replacing the paper version of the IFRB's International Frequency List by microfiche in 1983. Each member administration will be supplied by the ITU with two microfiche readers with the first microfiche edition of the list.[12]

Despite the problem of rising costs, the ITU continues to mount a major publication effort. In 1979, for example, the ITU produced 777,020 copies of some 466 publications at a cost of 10,684,278.47 Swiss francs.[13]While this does not bring the ITU anywhere close to the output of the United Nations or one of the richer Specialized Agencies such as UNESCO, it does establish the ITU as an important publisher among the UN related agencies and one of the major publishers in the field of telecommunication.

TELECOM

One other informational activity of the ITU should be discussed in some detail before we conclude this section, and that is the World Telecommunication Exhibition (TELECOM) and the activities that surround it. In the late 1960s, somewhere in the ITU Secretariat, the idea was conceived of holding an exhibition of telecommunications technology in Geneva at the same time as major ITU administrative radio conferences. The first such exhibition was organized in Geneva in conjunction with the 1971 World Administrative Radio Conference for Space Communications. While organized by the ITU and sponsored by the Swiss Government, the practical side of the affair was entrusted to OEGEXPO, "a nonprofit-making foundation

---

[12] See U.S., Report of the U.S. Mission in Geneva to the Secretary of State on the 36th *Session of the Administrative Council, Geneva, 1-19 June 1981,* Telegram No. 06995. The International Frequency List has been available on magnetic tape for a number of years, which for some administrations resulted in a substantive savings in both time and money.

[13]See ITU, *Report on the Activities of the International Telecommunication Union in 1979,* Geneva, 1980, pp. 133-137.

established for the express purpose of organizing exhibitions in Geneva, particularly for the United Nations Organization and its specialized agencies."[14] Some 250 exhibitors from fourteen countries took part in TELECOM 71.

The Secretary-General reported to the 1972 Administrative Council that TELECOM 71 had been a complete success. It had not cost the ITU any money; indeed, it has resulted in a 50,000 Swiss franc surplus. The Secretary-General further reported that he had consulted with member administrations concerning holding another such exhibition in 1974 and that "a majority of administrations were in favor of holding another exhibition soon."[15] A discussion followed dealing with two issues: whether the ITU was authorized to hold such meetings and whether the next one should be held in 1974 or later. The Council finally decided to authorize the Secretary-General to make a more accurate consultation with member administrations on whether they were interested in having another exhibition in 1974 and to postpone the final decision to the Council's next meeting.

The issue was debated at some length at the Council's 1973 session, when faced with an extensive report on the matter from the Secretary-General and a proposal to hold the next TELECOM in 1974. As in the previous session, there was a confusion over issues. One group of councilors supported both the concept and call for another exhibition in 1974. Another group composed of representatives of both developed and lesser developed countries lent support to the concept of having such exhibitions but questioned the need for one as early as 1974. An additional group, also mixed, opposed the whole concept, arguing that the ITU should not be involved in a commercial enterprise such as that represented in the TELECOMs.[16]

After a lengthy discussion, the proposal to organize a TELECOM exhibition in 1974 failed on a vote of eight in favor, ten in opposition, and six abstentions. A further discussion ensued in which some councilors stated that their vote dealt only with the propriety of

---

[14]See ITU, *Geneva: World Telecommunication Exhibition,* Geneva, 1976, p. 2.

[15]ITU, Administrative Council, 27th Session, Geneva, May-June, 1972, Document No. 4402-E (CA27-144) mimeo, p.2.

[16]ITU, Administrative Council, 28th Session, Geneva, April-May, 1973, Document No. 4524-E (CA28-108), pp. 4-6.

holding a TELECOM in 1974 and others stating that they had voted on the principle involved. Discussion was finally ended at the 1973 Administrative Council meeting by a seldom used motion for closure of debate which was passed by thirteen to eight, with one abstention.[17]

Between the end of the 1973 Council session and the beginning of the 1973 Plenipotentiary, the Secretary-General picked up some additional support. A "draft opinion" was introduced at the 1973 Plenipotentiary by twenty-four delegations, mostly from the lesser developed countries, approving of the holding of such exhibitions under the auspices of the ITU as long as there would be no cost to the ITU and no commercial interests would be involved.

Speaking in support of the opinion, the delegates of Saudi Arabia, Nigeria, India, and the Ivory Coast felt that "such exhibitions were of the utmost value to the developing countries as officials from their administrations, when attending a conference, had an opportunity to get up-to-date information on the latest advances in telecommunication techniques and see the most recent equipment on display under one roof. That saved them a great deal of time and traveling and obviated the need to visit manufacturers individually when acquiring equipment." Another group of countries, including the United Kingdom, the U.S., Australia, and Argentina, argued that no matter how one looked at TELECOM it was a commercial enterprise involving the sale of equipment. The ITU, it was argued, should not "assist equipment manufacturers to sell their goods and I.T.U personnel should not be diverted for the purpose of servicing such exhibitions . . . "[18] The opinion (Opinion No. 3) approving of the holding of such exhibitions under the auspices of the ITU, "provided that this involves for the Union no charge on its budget and no commercial interest," was passed by the underwhelming vote of 36 in favor, 31 opposed, and 30 abstentions.

The second TELECOM exhibition was held in Geneva in 1975 in conjunction with the Second Session of the Regional Administrative Conference for LF/MF Broadcasting. There were 350 exhibitors

---

[17]See ITU, Administrative Council, 28th Session, Geneva, April-May, 1973, Document No. 4525-E (CA28-109), pp. 2-3.

[18]ITU, *Minutes of the Plenipotentiary Conference of the International Telecommunication Union, Malaga-Torremolinos, 1973*, Geneva, 1974, pp. 479-480.

in 1975, and the profit was 2,534,084.04 Swiss francs. The third TELECOM, held in Geneva at the beginning of the World Administrative Radio Conference of 1979, had a number of additional events, including a book fair and a World Telecommunications Forum under the chairmanship of the Secretary-General of the ITU, in which a number of representatives from the ITU, administrations, and private operating agencies shared their thoughts on various aspects of the telecommunications revolution. The main ingredient remained displays of equipment, however. The number of exhibitors jumped from 350 in 1975 to 600 in 1979, but the surplus was only 74,612.88 Swiss francs.

The fourth TELECOM is scheduled for 1983 in conjunction with the Second Session of the Conference for the Planning of FM Sound Broadcasting in the band 87.5-108 MHz. Although it will be held in the new, much larger Geneva Exposition Palace, at the time of writing it was already completely subscribed.

Only one discussion has been held on the subject in the Administrative Council since 1973. At that meeting in 1980, the Secretary-General was requested to look into the possibility of holding TELECOMs in countries other than Switzerland, despite the Secretary-General's argument that he could not ensure the "impartiality" of exhibitions held outside of Geneva. The Secretary-General was also requested to investigate a "more methodological charging of Union costs to TELECOM accounts (for which the Union is reimbursed) including some percentage of his own time."[19]

Since the accounts of the TELECOMs do not appear in the Annual Report of Activities of the ITU or any other official ITU document available to the public, it is difficult for the outsider to determine the actual relationship of the TELECOMs to the ITU. A request to the ITU for information on this point elicited a response to the effect that the TELECOM concept is linked to the ITU "by approval at the 1971 Administrative Council session of the exhibition TELECOM 71" and by adoption of Opinion No. 3 by 1973 Plenipotentiary Conference at Malaga-Torremolinos that was mentioned above. Further, concerning financial arrangements, as stated in Opinion No. 3, ITU has no financial involvement in the running of TELECOM, which is

---

[19]See *Report of the U.S. Mission in Geneva to Secretary of State, International Telecommunication Union,* 35th Session of the Administrative Council, Incoming Telegram No. 07768, Washington, D.C., 1980.

a self-supporting activity. Staff members working for TELECOM are specially recruited and paid by the TELECOM budget and all other facilities such as offices, postal and telex services, are also charged to this budget. The ITU auditors check the TELECOM account and the Administrative Council is informed of the results. Cost per square meter is calculated to cover expenses. In 1971, 1975, as well as 1979, the TELECOM accounts showed a small positive balance which has been used every time for the promotion of the next exhibition. The fourth exhibition will take place in 1984, and as of now, most of the space is already committed.[20]

Despite some remaining confusion concerning the relationship of TELECOM to the ITU, as exemplified in the Administrative Council's request to the Secretary-General that he look into a more methodological charging of the Union's costs to TELECOM, the decision of the Malaga-Torremolinos Plenipotentiary, as close as it was, seems to have quelled any open opposition to the TELECOM concept. The support of the developing countries was the deciding factor. As long as it has that support, and it does not become an additional expense to the ITU's members, TELECOM exhibitions will undoubtedly continue to be a legitimate, new informational activity of the ITU.

ITU Documentation Center

In their search for additional ways to assist the administrations from developing countries in improving their domestic telecommunications networks, the delegates of the 1973 Plenipotentiary decided to give them better access to telecommunication documentation.

The ITU has always had a mission to make advances in telecommunication technology and new techniques available to administrations. Indeed, the Convention directs the Secretary-General to "collect and publish, in suitable form, data, both national and international, regarding telecommunications throughout the world."[21]

---

[20]From a letter 25, March 1980, from R. Fontaine, Chief, Public Relations Division, ITU General Secretariat. A subsequent letter from Fontaine dated July 3, 1981 contained information concerning finances that was used in this section.

[21] ITU *International Telecommunication Convention,* Malaga-Torremolinos *1973,* Art. 56, p. 1.

The ITU also has an excellent, small library in the older part of the ITU headquarters building, which is manned by an efficient and helpful staff. This library has a complete collection of ITU documents, both public and restricted, and a good selection of books and journals dealing with telecommunication matters. There are even a number of government documents from some of the member administrations. For the staff of the ITU and for those who are fortunate enough to be able to visit the Geneva headquarters, the ITU library is a valuable information resource.

The 1973 Plenipotentiary took an additional step in this direction. In view of the vast amount of valuable information in the government and international document collections throughout the world and in view of its potential value, especially to the developing countries, it was agreed in Resolution Number 47 to direct the Secretary-General, with the assistance of other permanent organs, to study the "creation of a documentation and bibliographical reference center for telecommunications to be responsible for:

1. Facilitating the use of documentation published by the Union;

2. Collaborating with other international or national documentation centers in the exchange of bibliographical references in order to avoid duplication of work, reduce expenses and, at the same time, centralize world information on telecommunications;

3. Placing this information at the disposal of Members and officials and experts of the Union."[22]

The results of a six year study on the matter were reported to the ITU membership on 21 August 1981 by the working group that has been established within the ITU for that purpose. The report included an identification of potential users, the volume of documents to be indexed, the indexing and retrieval parameters, and the facilities that would be needed. The report concluded that in view of its findings and the affirmative response "not only of the majority of Administrations of Member countries of the Union but also of the scientific and industrial organizations," it had come to the conclusion that "the Union can only benefit from the creation of this new facility."[23]

---

[22] ITU, *International Telecommunication Convention, Malaga-Torrremolinos, 1973.*
[23] See ITU, General Secretariat Document No. 105 (Rev. 2)-E, "Report of the Working Party 'Documentation Centre'," contained in General Secretariat,Circular Letter No. 426, INF/BIBL/DOC 24, 21 August 1981, p. 37.

Conclusions

The information function of the ITU can easily be overlooked, but ultimately it is one of the most important. In this effort the ITU acts as a central collection point for massive amounts of information-concerning telecommunications throughout the world which would be extremely difficult, if not impossible, to obtain by any other means. This information, which is available to all of the members of the Union, is an essential element in the planning and operation of the world's domestic and international telecommunication systems.

IV

Problems
and Prospects

chapter 14

CONCLUSIONS

If the major function of an international organization is to provide its members with services which they find difficult or impossible to provide for themselves, then the existence of the International Telecommunication Union is justified indeed. By its very nature, international telecommunication demands the cooperative efforts of nations. In many respects,even the successful operation of domestic telecommunication networks requires similar cooperative efforts. Few nations could easily "go it alone." As a reflection of this fact, the membership of the ITU is one of the largest of the entire United Nations family of Specialized Agencies. Joining the ITU has always been a priority item on the agenda of any newly independent state.

For an international organization to endure, however, there is an additional requirement: the ability to adapt to changes in the economic and political climate in which it operates and to changes in the technology that it regulates. That the ITU has flourished for 116 years despite two world wars and many lesser conflicts is ample evidence that the ITU has had this ability in the past. Only one other international organization, the Universal Postal Union, has come close to the ITU in longevity. Not too surprisingly, it is also involved in the regulation of a form of international communication.

The Past as Prelude

In the beginning, the International Telegraph Union provided a forum in which the more advanced telegraph administrations of Europe could work toward the establishment of a regional telegraph network for the benefit of the governments involved, commercial interests, and others. This was accomplished in a relatively short time, and the Union and the network it served grew rapidly as more countries saw the advantages of the electric telegraph. As non-European telegraph administrations became members, the International Telegraph Union rapidly became truly international.

The technological breakthrough of the telephone was easily incorporated into the mandate of the ITU because, in the beginning, it was considered no more than an adjunct to the telegraph, thus not a threat to the entrenched interests of telegraph bureaucracies. While it was also a wire service, thus similar in many ways to the telegraph, the state of its technology and the language barriers involved resulted in its being used primarily for local purposes. For a long time the telegraph administrations involved were happy to keep the status quo. The few additional rules necessary for this new service were quickly and easily added to the telegraph regulations.

Radio posed new problems when it appeared on the scene at the turn of the century. Like the means of communication that had preceded it, one of the first prerequisites for its successful use was the interconnection of systems. Restrictive practices, such as those introduced by the Marconi Company, could only retard the creation of an international radio network to provide for the needs of ships at sea, the one area where the telegraph and telephone could not serve. The group of administrations that met to solve this and related problems of the new wireless telegraph decided to create a new entity, complete with a separate convention and service regulations. The new group did, however, entrust the Bureau of the Telegraph Union with the job of providing administrative support. This hybrid scheme was devised because of the unique characteristics of radiocommunication and a desire to have the active participation of the United States, which had boycotted the International Telegraph Union. In the U.S., there existed a form of private-sector telegraph and telephone operations. It was soon realized that maintaining two different sets of arrangements in the field of telecommunication was inappropriate, and a merger was effected in 1934, which created the International Telecommunication Union.

The only major change in the structure of the ITU in the years before World War II was the addition of the consultative committees to the periodic conferences and the creation of the Berne Bureau. The conferences were having an increasingly difficult time carrying out their work load, due to the increasing complexity of the technology and the increase in members. In addition, there was a need to maintain a continuing dialogue for the exchange of information on new technology and developing standards. The three international consultative committees were created to assist in these areas.

The exceptionally rapid advances in radiocommunication in the 1930s and 1940s brought the fledgling Union to a major crisis. Many of the radio frequency bands had become so crowded with stations that it was felt that the future of some radio uses were in jeopardy. In addition, by the end of World War II, a number of the more influential members of the ITU had become convinced that the Berne Frequency List no longer reflected an equitable distribution of radio channels among the nations of the world. Thus, a radical new system of rights vesting was necessary. After much deliberation and at the prodding of the United States, the delegates to the Atlantic City conferences of 1947 decided to adopt a variation on the international planning approaches originally discussed after the First World War and implemented at the regional level over the next two decades. Using modern engineering methods, a new "engineered" international frequency list would be created to meet the needs of all nations, and an International Frequency Registration Board would thereafter maintain it in the light of future needs.

It was much easier to adapt the ITU to the new post-World War II environment as exemplified by the new United Nations organization. The Secretariat was taken out of the custody of the Swiss government and internationalized. A limited membership Administrative Council was created to provide a liaison with the new United Nations, to coordinate the work of the permanent organs, and to take the place of the Swiss government in the supervision of the Secretariat in its activities between plenipotentiary conferences.

Another crisis occurred in the late 1940s and 1950s when it was realized that it would not be possible to create the new international frequency list on which it was thought the future rational use of radio communication depended. In a few uncontentious areas such as commercial aircraft frequencies, an allotment plan was achieved. But in the greater portion of the radio frequency spectrum, the plan reached an impasse. This was especially true for the highly contentious shortwave broadcasting bands. The ITU members reacted to the impasse by accepting the situation and, in areas where agreement was impossible, reverting to the former system of notification and registration through the new IFRB. For the shortwave broadcasting frequencies, the Board could do little more than assist ITU members in coordinating their seasonal schedules and noting conflicts. Also, the size of the Board was reduced to reflect a lesser role in international frequency management. Despite the fears of many, these new arrangements were adequate to meet the needs of the majority of members throughout this period.

New Challenges

The ITU is again faced with a number of important challenges with which it must deal in order to continue to serve the needs of its members. The most fundamental challenge involves the ITU's jurisdiction over its subject matter. Since 1934, the Union has enjoyed dominant jurisdiction over international arrangements concerning "telecommunication." That word, however, is rapidly losing its clarity of scope and meaning. It is becoming an integral, inextricable part of a larger information system. The world's knowledge and information is increasingly being captured, processed, stored and transported in electronic form. This is accomplished largely through computers which control these systems as extensions of their own "intelligence," much as the nervous system of the human body is an extension of its brain. Telecommunication, or "information transport" as it is being increasingly referred to, is losing some of its identity in the process. The change is so rapid that communications will probably be reshaped more in the next decade than it has in the past 140 years.

In this period of change, the international community urgently needs forums to work out the necessary arrangements, including guiding principles, strategies, and standards. A number of other organizations, each with its own expertise or bias, are attempting to carve out their own jurisdictions, as well as to honestly respond to international needs. On the technical side these include the International Organization for Standardization (ISO) and the International Federation for Information Processing (IFIP). On non-technical issues, they include UNESCO, the UN itself, the Intergovernmental Bureau for Informatics (IBI), and the Organization for Economic Cooperation ad Development (OECD). The fundamental question facing the ITU is not whether it will play a role, but rather how much of a role and what kind.

An additional contemporary challenge is, in many respects, a resurrection of the problem which faced the ITU in 1947. As certain frequency bands become increasingly crowded, there is a new demand for equitable rights-vesting arrangements. This includes arrangements for frequencies which have been fought over in the past, such as those which are used for high frequency broadcasting and, more recently, the frequencies most suitable for satellite communication, as well as the orbital positions of the satellites themselves. The problem today is compounded by the vastly increased number of countries involved and the inability of many of them to deal satisfactorily with the existing arrangements.

A third, interrelated challenge is that of reconciling the needs of the developing nation majority with those of the developed nation minority. This includes achieving a satisfactory solution to the problem of rights-vesting mechanisms, the amount of the contributions that the two groups should make to the expenses of the ITU, and what portion of its resources the Union should devote to assisting the poorest of its members in building domestic telecommunication facilities. This is potentially one of the most disruptive challenges facing the ITU today.

Regulatory Concerns

The ITU's most significant function is the regulation of the world's telecommunications. The term regulation must be used guardedly, however, because the Union itself does not have any explicit enforcement powers; it merely relies upon member states to see that the decisions of the ITU are carried out. However, the regulatory arrangements devised by the Union are not of a compulsory nature. Unlike many kinds of international acitivities, communication requires the cooperation of all the affected parties. For instance, two radio stations operating in an uncoordinated manner in the same areas are likely to diminish the value of each other's services. The states which provide transcontinental telephone cables must be in agreement before they can function. Even in postal communication, the cooperation of least two states is required if international mail is to reach its destination. Moreover, the economic and technical efficiency which results from the adoption of common designs and procedures creates a substantial inducement to conform to international agreements. These realities of communication tend to make the ITU's international regulations coercive in nature. This explains why the the oldest international organizations, the ITU and the Universal Postal Union, deal with communication, and why their regulatory product is closely adhered to by the world community.

The ITU's regulatory function can be divided into two generic categories: standards-making and rights-vesting. The first encompasses thousands of technical specifications and operating agreements which were devised to allow the common use of a telecommunication network or service and to effect a measure of compatibility among separate systems. Because information and communication technology is advancing at a rapid pace, the ITU will have to increase its standards-making work in the future. It will also face a crucial issue in standards-making: the effort to achieve a difficult and delicate balance between regulations which are explicit enough to promote the implementation of new telecommunication facilities and yet not so detailed and inflexible that they prevent significant innovation. While this balance is primarily determined by economics, it is also substantially affected by regulatory policy, national security, and technological considerations. The goal must be to achieve the greatest consensus possible, particularly among parties with the greatest stake in the outcome. It is a complicated process, one which may defy rational analysis.

The ITU's other basic regulatory function, rights-vesting, is less strongly influenced by technology. Indeed, the basic approach to this function has remained unchanged over a number of decades. Like standards-making, it calls for a careful balancing of interests. However, the number of participants in the ITU forums has increased significantly over the past few decades, and with the emergence of the Third World majority, the participants have increasingly divergent interests. This not only tends to prevent a consensus on many matters, but also has set new directions in the conferences, directions not always favored by the older, industrialized nations. This is especially apparent in conferences dealing with the use of satellites and those whose task it is to allot spectrum resources among member administrations. The next decade or two will be difficult ones, as individual administrations and user groups bring pressure on ITU forums for exclusive parcels of the increasingly valuable spectrum resources they believe essential to their particular needs.

Possibly, however, new technological developments will make many of these problems moot. For example, a computer-controlled, multiple-beam space platform capable of simultaneously providing services to many regional users, would reduce international competition over equitable access to the satellite orbit and radio frequency resources.

In the immediate future, a series of administrative radio conferences has been scheduled to fashion a consensus on arrangements for satellite communication and H.F. broadcasting frequency usage. If substantial agreement cannot be reached (not a rare situation in the past), the larger, relatively isolated, developed nations may decide to go their own way, making more reservations to the final acts of conferences and relying more on bilateral or regional arrangements. Such action would not only make international telecommunication increasingly difficult and expensive, but it might delay the development and implementation of large multi-user space facilities which would be most useful to the developing countries.

The ability of the ITU and its members to achieve a consensus on regulatory matters, especially those concerning rights-vesting, may largely determine whether the peoples of the world will enjoy the many fruits that modern telecommunication technology can bring.

Structural Concerns

For the most part, the ITU has all of the basic elements it needs to face the challenge of the future. There are conferences and meetings of delegates from member countries, a limited membership Administrative Council to oversee its activities between conferences, and a secretariat of international civil servants in Geneva to provide

support services. The cost of maintaining the Secretariat and conducting the conferences is borne in large part by contributions of the
member administrations, with some additional funds from private
operating agencies and scientific and industrial organizations
which participate in various organs, especially the CCIs.

It has been the tradition in the ITU, as in most international organizations, that all decisions which can have even a potential effect on
the vital interests of member states must be made in a conference to
which all affected administrations have the right to send delegates,
even though those decisions are not binding until they are ratified
by the governments of those same states.

The increased membership of the ITU and the increased complexity
of the technology is placing strains on the ITU's primitive conference system. Delegates to conferences find it increasingly difficult
to carry out the work that has been scheduled within the time limits
set by the organizers. This has led to the delegation of issues posing
even the slightest possibility of a controversy to working groups and
sub-working groups. The 1979 World Administrative Radio Conference was a good example of this trend. There is also a tendency to
divide subjects into ever smaller and more detailed categories, each
with its own conference or meeting. The result has been an increasing strain on member administrations, especially those which do
not have the necessary trained personnel or adequate finances,
producing a feeling of alienation on the part of countries without
those resources. There also seems to result a concentration on
details and the neglect of some of the issues with which an international organization such as the ITU should deal.

Another strain on the conference system is introduced by its general
inability to adapt to the rapid changes in technology. On the telegraph and telephone side, the ITU has effected a solution by shifting
virtually all the technical provisions out of the domain of the administrative conferences and Regulations into the CCITT and its
Recommendations. The radio side of the ITU, however, appears less
innovative and is fairly inflexible.

The Administrative Council seems to have lived up to most of the
expectations of its creators. In maintaining communications among
the UN organizations, in overseeing the work of the permanent

organs, and in working to keep them within budget limits, it has served member administrations well.

However, the Administrative Council is also suffering from some of the same ailments that plague conferences. As its size has increased, it is finding it difficult to carry out its assigned tasks in the time alloted to it. Most important, perhaps, it is so involved with details that, with few exceptions, it has not been able to help with the exploration of basic issues. To cope with those problems, suggestions have been made to hold more than one session a year, or to create a permanent executive committee. Any solution, however, runs up against the problem of how much authority any international organization should give to a body which consists of only a small portion of the total membership.

The foundation on which any international organization rests is its Secretariat. In the case of the ITU, it is composed of a General Secretariat headed by a Secretary-General and a Deputy Secretary-General, who are elected by the plenipotentiary, and the smaller, specialized secretariats of the CCIR, CCITT, and the IFRB. It is the Secretariat which is the heart of the organization: collecting and disseminating all kinds of information on telecommunications; providing help to administrations; maintaining the administrative mechanisms devised by the present and past ITU membership; administering technical assistance programs; and doing much of the work for the preparation and conduct of the constant round of conferences and meetings.

Because of the preponderant role that he plays in these activities, the Secretary-General is the most visible element in the ITU. This, combined with the length of the term of his office, makes him the principal symbol of the ITU in the minds of many who have dealings with it. In view of some of the problems afflicting the ITU's representative organs, in many respects the leadership role of the Secretary-General has become as real as it is symbolic.

One of the more unusual elements in the ITU is the consultative committee. Originally three in number but reduced to two in 1956, the CCIs are made up of plenary assemblies, a myriad of working groups and parties, specialized secretariats, and directors elected by their respective general assemblies. The uniqueness of the CCIs stems from the independence that they exercise within the ITU as regards both structure and function. It also stems from the fact that representatives of government telecommunication administrations work side by side with representatives of private operating agencies and various scientific and industrial organizations interested in the progress of telecommunication.

One of the original purposes of the consultative committee was to help speed up the work of conferences by providing a body which would study technical issues in the intervals between conferences. The CCIs still carry out this function, which in the case of radio is increasingly important as administrative radio conferences tend to proliferate. However, the CCI's most important function today is the preparation and adoption of international standards. As the electronic information world races toward integrated services digital networks, the CCIs are becoming a focal point for world agreement on strategies, principles, and specifications for virtually every aspect of their design and operation.

The work of the CCI at the forefront of technology, however, is not their only responsibility. Increasingly, the Third World nations have sought to have some of the CCI resources devoted to developing technological standards for equipment and systems appropriate for poor, largely rural populations. This has not been easily brought about because of the largely voluntary nature of CCI work and the fairly significant costs associated with participation. In addition, the participants in CCI work are naturally attracted toward new, advanced developments. Whether the work can be conducted so as to satisfy the needs of developing countries while maintaining the effectiveness of the committees without alienating those who have volunteered their services is a serious problem that needs resolution.

In any case, because the distinctions between radio, telegraph, and telephone are becoming increasingly irrelevant, it can be expected that the two remaining CCIs will be merged into one some time in the not too distant future. How soon this occurs will be determined by the interplay between the changing technology, which is providing increasing pressure for a merger, and the entrenched interests of government and private bureaucracies, which tend towards the maintenance of the status quo.

The most exotic organ in the ITU structure is the International Frequency Registration Board (IFRB), which is composed of five independent members who are thoroughly qualified by technical training in the field of radio and possess practical experience in the assignment and utilization of frequencies. Many international organizations have co-opted expert groups to help them carry out technical studies that they cannot carry out themselves; but few have made such groups permanent organs in their structure, and fewer still are elected by the whole membership and given their own specialized secretariat.

If the task of the IFRB were confined to the recording of frequency assignments and keeping the international frequency list up to date, even with the addition of duties concerning the publication of high frequency broadcasting schedules and the coordination of new satellite communication systems, then there might well be some question as to whether such an organ is justified. However, its attractiveness as a friend to the Third World nations which constitute the new majority seems to have assured it a continuing role within the ITU.

The very existence of the consultative committees and the IFRB, which have no equivalents in most other international organizations, has raised one of the most vexing of the ITU's internal problems, the problem of federalism. In the past, at least, it has been a maxim that for these bodies to carry out their functions effectively, it was necessary for them to have the help of a secretariat made up of specialists in their respective fields of work. Only the heads of these organs were considered competent enough to determine the exact type of expertise that was needed, and they were thus given a certain amount of authority over their specialized secretariats independent of the Secretary-General. With the unavoidable overlapping of functions among the permanent organs, combined with the ever-present possibility of personality conflicts, it was inevitable that friction should occur, which was the case almost from the beginning.

A number of attempts have been made over the years to solve this problem, including the creation of the intra-agency consultative committee, but none has been successful. Unless a forthcoming plenipotentiary should make some major, drastic changes in the ITU's structure or its functions, which is not likely, it is certain that the problem of federalism will continue to surface with some regularity to annoy all of those involved, especially the Administrative Council.

As complex as the structure and functions of the ITU are, its cost appears to be quite reasonable. The budget of the ITU has consistently been among the lowest of the Specialized Agencies; and, between the limits on expenses set by the plenipotentiary and the oversight function of the Administrative Council, there is little opportunity for overspending.

Perspectives

The ITU has a number of obvious strengths. It has a tradition of 116 years, which gives it a great deal of natural momentum. It has a structure which has adapted to the changing demands of new technology in the past without a great deal of difficulty. And it has the support of a number of dedicated individuals in the Secretariat, in the member administrations, and in private operating agencies and organizations which participate in ITU conferences and meetings. Few international organizations can match these resources.

The emerging communication revolution will demand even more of the ITU in the future, however. Its utility as a forum for the coordination of the information society of the future to the benefit of all concerned will depend on a number of factors. It will depend on the ability of the member administrations to devise a common strategy in a timely manner. It will depend on the astuteness of individuals in the Secretariat and in key delegations to conferences and meetings. It will depend, also, on whether a preponderance of the members of the ITU want to provide the necessary arrangements.

The ITU and its members will need all of the help that they can get in this arduous endeavour. The first and most obvious task is to eliminate those defects that exist in the ITU's present structure. Some affect the ability of the organization to carry out its technical work, while others are non-technical in the strictest sense of the word. There are none, however, which could not be remedied by its members with even a modicum of goodwill and understanding.

There is little doubt that the ITU has an important role to play in future communications. The question is, however, how much a role it will play and what kind. The answers lie in the decisions to be made in the many important conferences that will be held during the next decade.

Annexes

ANNEX A
Organizational Structure of the ITU*

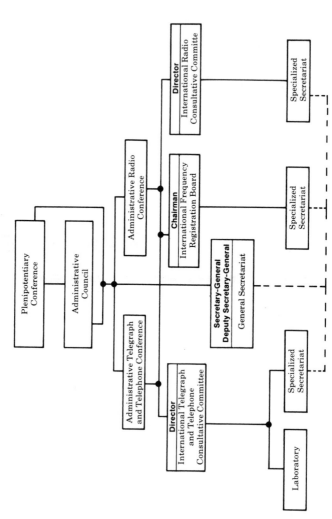

*The ITU Convention designates the CCIR, CCITT, IFRB, and the General Secretariat as the permanent organs of the ITU. The ITU also has a Coordination Committee, not shown in this diagram, made up of the Secretary General, Deputy Secretary General, Director of the CCIR, Director of the CCITT, and Chairman of the IFRB whose major task it is to "assist and advise" the Secretary General. See ITU, *International Telecommunication Union, Malaga-Torremolinos, 1973*, Arts. 5 and 12.

ANNEX B
List of Members of the ITU
with Indication of the Date of Their Admission in
the Union (1 January 1981)

Members	*Admission date*
Afghanistan (Democratic Republic of)	12. 4.1928
Albania (Socialist People's Republic of)	2. 6.1922
Algeria (Algerian Democratic and Popular Republic)	3. 5.1963
Germany (Federal Republic of)	1. 1.1866
Angola (People's Republic of)	13.10.1976
Saudi Arabia (Kingdom of)	7. 2.1949
Argentine Republic	1. 1.1889
Australia	27. 5.1978
Austria	1. 1.1866
Bahamas (Commonwealth of the)	19. 8.1974
Bahrain (State of)	1. 1.1975
Bangladesh (People's Republic of)	5. 9.1973
Barbados	16. 8.1967
Belgium	1. 1.1866
Benin (People's Republic of)	1. 1.1961
Byelorussian Soviet Socialist Republic	7. 5.1947
Burma (Socialist Republic of the Union of)	15. 9.1937
Bolivia	1. 6.1907
Botswana (Republic of)	2. 4.1968
Brazil (Federative Republic of)	4. 7.1877
Bulgaria (People's Republic of)	18. 9.1880
Burundi (Republic of)	16. 2.1963
Cameroon (United Republic of)	22.12.1960
Canada	22.12.1960
Cape Verde (Republic of)	10. 9.1976
Central African Republic	2.12.1960
Chile	1. 1.1908
China (People's Republic of)	1. 9.1920(1)
Cyprus (Republic of)	24. 4.1961
Vatican City State	1. 6.1929 (1866*)
Colombia (Republic of)	24. 4.1961
Comoros (Federal and Islamic Republic of the)	6. 1.1976
Congo (People's Republic of the)	13.12.1960
Korea (Republic of)	31. 1.1952
Costa Rica	13. 9.1932
Ivory Coast (Republic of the)	23.12.1960

1) In June 1972, Administrative Council Resolution 693 restored all rights to the People's Republic of China in the ITU.
* Church states.

Members	*Admission date*
Cuba	16. 1.1918
Denmark	1. 1.1866
Djibouti (Republic of)	22.11.1977
Dominican Republic	11.7 7.1926
Egypt (Arab Republic of)	9.12.1876
El Salvador (Republic of)	12.10.1927
United Arab Emirates	27. 6.1972
Ecuador	17. 4.1920
Spain	1. 1.1866 (2.5.1951)
United States of America	1. 7.1908
Ethiopia	20. 2.1932
Fiji	5. 5.1971
Finland	1. 9.1920
France	1. 1.1866
Gabon Republic	28.12.1960
Gambia (Republic of the)	27. 5.1974
Ghana	17. 5.1957
Greece	1. 1.1866
Guatemala (Republic of)	10. 7.1914
Guinea (People's Revolutionary Republic of)	9. 3.1959
Guinea-Bissau (Republic of)	15. 1.1976
Equatorial Guinea (Republic of)	2. 7.1970
Guyana	8. 3.1967
Haiti (Republic of)	10.10.1927
Upper Volta (Republic of)	16. 1.1962
Honduras (Republic of)	27.10.1925
Hungarian People's Republic	1. 1.1866
India (Republic of)	1. 1.1869
Indonesia (Republic of)	1. 1.1949
Iran (Islamic Republic of)	1. 1.1869
Iraq (Republic of)	12.11.1928
Ireland	8.12.1923
Iceland	1.10.1906
Israel (State of)	24. 6.1948
Italy	1. 1.1866
Jamaica	18. 1. 1963
Japan	29. 1.1879
Jordan (Hashemite Kingdom of)	20. 5.1947
Democratic Kampuchea	10. 4.1952
Kenya (Republic of)	11. 4.1964
Kuwait (State of)	14. 8.1959
Lao People's Democratic Republic	3. 4.1952
Lesotho (Kingdom of)	26. 5.1967
Lebanon	12. 1. 1924
Liberia (Republic of)	10.10.1927
Libya (Socialist People's Libyan Arab Jamahiriya)	3. 2.1953
Liechtenstein (Principality of)	25. 7.1963

Members	*Admission date*
Luxembourg	2. 3.1866
Madagascar (Democratic Republic of	11. 5.1961
Malaysia	3. 2.1958
Malawi	19. 2.1965
Maldives (Republic of)	28. 1. 1967
Mali (Republic of)	21.10.1960
Malta (Republic of)	1. 1.1965
Morocco (Kingdom of)	1.11.1956
Mauritius	30. 8.1969
Mexico	1. 7.1908
Monaco	1. 7.1908
Mongolian People's Republic	27. 8.1964
Mozambique (People's Republic of)	4.11.1975
Nauru (Republic of)	10. 6.1969
Nepal	5.12.1957
Nicaragua	12. 5.1926
Niger (Republic of)	14.11.1960
Nigeria (Federal Republic of)	11. 4.1961
Norway	1. 1.1866
New Zealand	3. 6.1878
Oman (Sultanate of)	28. 4.1972
Uganda (Republic of)	8. 3.1963
Pakistan (Islamic Republic of)	26. 8.1947
Panama (Republic of)	14. 7.1914
Papua New Guinea	31.10.1975
Paraguay (Republic of)	27. 9.1927
Netherlands (Kingdom of the)	1. 1.1866
Peru	12. 7.1915
Philippines (Republic of the)	25. 5.1912
Poland (People's Republic of)	1. 1.1921
Portugal	1. 1.1866
Qatar (State of)	27. 1.1924
Syrian Arab Republic	12. 3.1973
German Democratic Republic	3. 4.1973
Democratic People's Republic of Korea	24. 9.1975
Ukrainian Soviet Socialist Republic	7. 5.1947
Roumania (Socialist Republic of)	9. 1.1866
United Kingdom of Great Britain and Northern Ireland	24.2.1871
Rwanda (Republic of)	12.12.1962
San Marino (Republic of)	25. 3.1977
Sao Tome and Principe (Democratic Republic of)	1. 9.1976
Senegal (Republic of)	15.11.1960
Sierra Leone	30.12.1961
Singapore (Republic of)	22.10.1965
Somali Democratic Republic	28. 9.1962
Sudan (Democratic Republic of the)	23.10.1957
Sri Lanka (Democratic Socialist Republic of)	1. 1.1897

Members	*Admission date*
South Africa (Republic of)	1.10.1910
Sweden	1. 1.1866
Switzerland (Confederation of)	1. 1.1866
Suriname (Republic of)	15. 7.1976
Swaziland (Kingdom of)	11.11.1970
Tanzania (United Republic of)	31.10.1962
Chad (Republic of the)	25.11.1960
Czechoslovak Socialist Republic	10. 1.1920
Thailand	21. 4.1883
Togolese Republic	14. 9.1961
Tonga (Kingdom of)	7. 1.1972
Trinidad and Tobago	6. 3.1965
Tunisia	14.12.1956
Turkey	1. 1.1866
Union of Soviet Socialist Republics	1. 1.1866
Uruguay (Oriental Republic of)	1. 7.1902
Venezuela (Republic of)	13. 8.1920
Viet Nam (Socialist Republic of)*	24. 9.1951
Yemen Arab Republic	1. 1.1931
Yemen (People's Democratic Republic of)	15. 8.1968
Yugoslavia (Socialist Federal Republic of)	9. 1.1866
Zaire (Republic of)	6.12.1961
Zambia (Republic of)	23. 8.1965

---

* By letter received by the General Secretariat on 2 November 1976, the Government of the Socialist Republic of Viet Nam has declared that State to be the continuator of the Republic of South Viet Nam so far as membership of the ITU is concerned.

ANNEX C
Participation in CCIR Working Groups, 1980

Administration	1	2	3	4	5	6	7	8	9	10	11	CMTT	CMV	TOTAL
Germany (Fed. Rep. of)	×	×	×	×	×	×	×	×	×	×	×	×	×	13
Saudi Arabia				×	×			×		×	×	×		6
Argentina		×			×									2
Australia	×	×	×	×	×	×	×	×	×	×	×	×	×	13
Austria	×				×	×			×	×	×	×		7
Belgium								×		×	×	×		4
Brazil				×				×	×	×	×		×	6
Bulgaria	×	×			×	×	×	×		×	×	×	×	10
Cameroun									×					1
Canada	×	×	×	×	×	×	×	×	×	×	×	×	×	13
China	×	×	×	×	×	×	×	×	×	×	×	×		12
Cyprus								×						1
Korea (Rep. of)	×		×	×	×	×			×					6
Cuba	×				×	×								3
Denmark				×	×			×	×	×	×	×		7
Ecuador								×						1
Spain	×			×	×	×	×	×		×	×	×	×	10
United States	×	×	×	×	×	×	×	×	×	×	×	×	×	13
Finland	×	×		×	×	×	×	×	×	×	×	×		11
France	×	×	×	×	×	×	×	×	×	×	×	×	×	13
Honduras								×						1
Hungarian (P.R.)	×			×	×	×		×	×	×	×	×		9
India	×	×	×	×	×	×	×		×	×	×	×		11
Indonesia				×					×					2
Iraq								×						1
Ireland										×	×	×		3
Italy	×	×	×	×	×	×	×	×	×	×	×	×	×	13
Japan	×	×	×	×	×	×	×	×	×	×	×	×	×	13
Nigeria	×		×	×	×			×	×					6
Norway	×			×	×	×		×	×	×	×	×		9
New Zealand	×		×	×	×	×			×	×	×	×		9
Panama								×						1
Papua New Guinea	×		×	×					×				×	5
Netherlands	×	×		×	×		×	×	×	×	×	×	×	11
Poland	×			×				×		×	×	×		6
Portugal	×	×	×	×	×	×	×		×				×	9
German Democratic Republic	×		×	×				×	×	×	×	×	×	10
Romania					×	×								2
United Kingdom	×	×	×	×	×	×	×	×	×	×	×	×	×	13
Sweden	×		×	×	×	×	×	×	×	×	×	×		11
Switzerland	×			×	×		×	×	×	×	×	×	×	10
Turkey										×	×			2
U.S.S.R.	×	×	×	×	×	×	×	×	×	×	×	×	×	13
Venezuela	×				×	×								3
Yugoslavia	×			×	×	×		×	×	×	×	×		9
Totals: 45	30	16	18	29	30	27	19	31	29	30	30	28	17	

*From ITU, Administrative Council, 36th Session, Geneva, June 1981, Addendum No. 5 to Document No. 5621-E, 1 May 1981, pp. 5.3/11-5.3/12.

Recognized private operating, agencies, etc.	Study Group													
	1	2	3	4	5	6	7	8	9	10	11	CMTT	CMV	TOTAL
Recognized private operating agencies	14	3	7	20	18	9	1	13	10	14	18	19	7	
International organizations	3	2	–	2	4	3	3	5	1	4	4	3	–	
Specialized agencies of the United Nations	1	1	–	–	–	–	–	3	–	–	–	–	1	
Scientific and industrial organizations	3	–	–	7	–	1	1	6	11	4	3	7	2	
Other organs of the ITU	2	2	2	2	2	2	2	3	2	2	2	3	3	
Registration	121	69	61	164	123	96	60	201	145	195	233	171	57	
Contributions	51	36	24	156	122	81	43	189	114	84	139	119	25	1097
										68				
										Joint 10-11				

ANNEX D*
Participation in CCITT Meetings, 1980

COUNTRIES (Administrations or Recognized Private Operating Agencies)	I	II	III	IV	V	VI	VII	VIII	IX	X	XI	XII	XIV	XV	XVI	XVII	XVIII	CMBD	World	Africa	Latin America	LTG	GAS3	GAS5	GAS6	VII Plenary Assembly	TOTAL
																			(Plan)	(Plan)	(Plan)	(Joint Working Party)	(Special Autonomous Study Groups)				
Afghanistan																											—
Albania																										x	1
Algeria			x																	x					x	x	4
Germany (Fed. Rep. of)	x		x	x	x	x	x	x	x	x	x	x	x	x	x	x	x	x	x						x	x	20
Angola																											—
Saudi Arabia								x																	x	x	3
Argentina	x	x	x					x													x					x	6
Australia	x	x		x	x	x		x	x	x	x	x	x	x	x	x	x									x	16
Austria			x	x	x	x	x						x	x	x	x	x									x	11
Bahamas																											—
Bahrain																			x								1
Bangladesh																											—
Barbados			x																								1
Belegium	x		x	x			x		x	x	x	x	x	x	x	x	x		x						x	x	16
Benin			x																x							x	3
Byelorussia																										x	1
Burma																											—
Bolivia																											—
Botswana																											—
Brazil			x							x	x	x		x		x	x						x				8
Bulgaria											x	x			x	x											4
Burundi																											—
Cameroon			x																	x						x	3
Canada	x	x	x	x	x	x	x	x	x	x	x	x	x	x	x	x	x	x	x			x	x	x	x	x	24
Cape Verde																											—
Central African Rep.																				x							1
Chile																			x		x					x	3
China		x	x	x	x	x	x	x	x	x	x	x	x	x	x	x	x	x	x			x		x		x	21
Cyprus		x	x								x															x	4

*From ITU, Administrative Council, 36th Session, Geneva, June 1981, Document No. 5621-E (CA36-40), March 1981, pp. 5.4/27-5.4/32.

COUNTRIES (Administrations or Recognized Private Operating Agencies)	I	II	III	IV	V	VI	VII	VIII	IX	X	XI	XII	XIV	XV	XVI	XVII	XVIII	CMBD	World	Africa	Latin America	LTG	GAS3	GAS5	GAS6	VII Plenary Assembly	TOTAL
																			(Plan)	(Plan)	(Plan)	(Joint Working Party)	(Special Autonomous Study Groups)				
Vatican																					x						1
Colombia																										x	2
Comoros																				x							1
Congo			x																x							x	3
Korea (Rep. of)								x											x							x	3
Costa Rica																										x	1
Ivory Coast																			x	x						x	3
Cuba																			x							x	2
Djibouti																											1
Denmark	x	x	x	x	x	x	x		x	x	x	x	x	x	x		x		x					x		x	17
Dominican Republic																x							x				1
Egypt															x		x									x	2
El Salvador																											1
United Arab Emirates																											1
Ecuador																											1
Spain	x	x	x	x	x	x	x	x	x	x	x	x	x	x	x	x	x	x	x			x	x	x	x	x	23
United States of America	x	x	x	x	x	x	x	x	x	x	x	x	x	x	x	x	x	x	x			x	x	x	x	x	24
Ethiopia											x					x	x										2
Fiji																											1
Finland	x	x	x	x	x	x	x	x	x	x	x	x	x	x	x	x	x		x							x	17
France	x	x	x	x	x	x	x	x	x	x	x	x	x	x	x	x	x	x	x			x	x	x	x	x	24
Gabon																										x	3
Gambia																											1
Ghana																											1
Greece											x																3
Guatemala																											1
Guinea																			x							x	2
Guinea-Bissau																											1
Equatorial Guinea																											1
Guyana									x																		1
Haiti																											
Upper Volta																			x							x	3
Honduras																										x	2
Hungary	x	x	x	x	x	x	x				x	x			x	x	x	x	x						x	x	1
India	x	x	x	x	x	x	x	x	x		x	x	x	x	x	x	x	x	x					x		x	19

COUNTRIES (Administrations or Recognized Private Operating Agencies)	Study Groups and their Working Parties																		Plan			Joint Working Party	Special Autonomous Study Groups			VII Plenary Assembly	TOTAL
	I	II	III	IV	V	VI	VII	VIII	IX	X	XI	XII	XIV	XV	XVI	XVII	XVIII	CMBD	World	Africa	Latin America	LTG	GAS3	GAS5	GAS6		
Indonesia																			×					×	×	×	4
Iran																			×						×		2
Iraq											×								×								2
Ireland		×		×	×		×		×	×	×	×	×	×		×	×		×				×		×	×	16
Iceland																										×	1
Israel																								×			1
Italy	×	×	×	×	×	×	×	×	×	×	×	×	×	×	×	×	×	×	×	×	×	×	×	×	×	×	26
Jamaica																											—
Japan	×	×	×	×	×	×	×	×	×	×	×	×	×	×	×	×	×	×	×			×	×	×	×	×	24
Jordan																			×					×		×	3
Democratic Kampuchea																											—
Kenya																			×	×						×	3
Kuwait											×								×							×	3
Lao P.D.R.																											—
Lesotho																											—
Lebanon		×																	×							×	3
Liberia																										×	1
Libya																											—
Liechtenstein																											—
Luxembourg																										×	1
Madagascar																										×	1
Malaysia																			×							×	2
Malawi																											—
Maldives																											—
Mali			×													×			×							×	4
Malta																											—
Morocco																			×							×	2
Mauritius																											—
Mauritania																										×	1
Mexico																			×		×					×	3
Monaco																										×	1
Mongolia																											—
Mozambique																											—
Nauru																											—
Nepal																											—

COUNTRIES (Administrations or Recognized Private Operating Agencies)	I	II	III	IV	V	VI	VII	VIII	IX	X	XI	XII	XIV	XV	XVI	XVII	XVIII	CMBD	World	Africa	Latin America	LTG	GAS3	GAS5	GAS6	VII Plenary Assembly	TOTAL
Nicaragua																											—
Niger		x																									1
Nigeria	x	x	x							x	x								x						x	x	8
Norway	x	x	x	x	x	x	x	x						x	x	x	x	x	x			x	x	x	x	x	19
New Zealand			x				x	x											x	x						x	6
Oman																			x								1
Uganda											x								x				x		x	x	5
Pakistan																			x							x	2
Panama																					x						1
Papua New Guinea	x																										—
Paraguay	x																				x						2
Netherlands	x	x	x	x	x	x	x	x	x		x	x	x	x	x	x	x	x	x			x	x	x	x	x	23
Peru			x											x													2
Philippines			x																x								2
Poland			x	x						x	x															x	5
Portugal	x	x								x	x			x		x	x		x					x		x	10
Qatar										x	x																2
Syria														x													1
German Dem. Rep.	x	x	x			x	x				x	x			x			x	x							x	11
D.P.R. of Korea		x	x																								2
Ukraine																											—
Romania																			x							x	2
United Kingdom	x	x	x	x	x	x	x	x	x	x	x	x	x	x	x	x	x	x	x			x	x		x	x	24
Rwanda																			x							x	2
San Marino																											—
San Tome and Principe																											—
Senegal																				x				x	x	x	4
Sierra Leone																											—
Singapore																											—
Somalia																											—
Sudan		x									x																5
Sri Lanka																											—
South Africa (Rep. of)											x									x							2
Sweden	x	x	x	x	x	x	x	x	x	x	x	x	x	x	x	x	x	x	x			x	x			x	22
Switzerland	x	x	x	x	x	x	x	x	x	x	x	x	x	x	x	x	x	x	x			x			x	x	22

COUNTRIES (Administrations or Recognized Private Operating Agencies)	Study Groups and their Working Parties																		Plan			Joint Working Party	Special Autonomous Study Groups			VII Plenary Assembly	TOTAL
	I	II	III	IV	V	VI	VII	VIII	IX	X	XI	XII	XIV	XV	XVI	XVII	XVIII	CMBD	World	Africa	Latin America	LTG	GAS3	GAS5	GAS6		
Suriname																			x								1
Swaziland																			x					x		x	6
Tanzania	x																										1
Chad												x			x				x							x	3
Czechoslovakia									x	x									x							x	4
Thailand									x										x							x	3
Togo			x																								1
Tonga																											—
Trinidad and Tobago																									x	x	3
Tunisia																			x							x	2
Turkey																			x			x				x	2
U.S.S.R.	x	x	x	x	x	x	x	x	x	x	x	x	x	x	x		x	x	x								20
Uruguay					x	x															x					x	4
Venezuela					x	x					x								x								3
Viet Nam																											1
Yemen A.R.																											1
Yemen (P.D.R. of)																x											—
Yugoslavia						x	x												x							x	6
Zaire																			x								—
Zambia																			x								1
Total of Countries	25	27	30	21	24	21	23	19	19	20	33	23	19	26	20	24	25	17	76	11	9	13	14	17	20	85	

ANNEX E*
List of Recognized Private Operating Agencies, Scientific or Industrial Organizations, and International Organizations Taking Part in the Work of the CCIs.
(Position on 1 January 1980)

I. Recognized Private Operating Agencies

Nom - *Name* - Nombre	CCIR	CCITT
ALLEMAGNE (République fédérale d') - *GERMANY (Federal Republic of)* - ALEMANIA (República Federal de)		
Deutsch-Atlantische Telegraphengesellschaft..........	-	½
Deutsche Welle.......................................	½	-
Norddeutscher Rundfunk (NDR).........................	½	-
Zweites Deutsches Fernsehen (ZDF)....................	½	-
AUTRICHE - *AUSTRIA* - AUSTRIA		
Osterreichischer Rundfunk............................	½	-
Radio-Austria A.G....................................	½	½
BRESIL - *BRAZIL* - BRASIL		
Emprêsa Brasileira de Telecomunicaçoes (EMBRATEL)....	1	1
CANADA - *CANADA* - CANADA		
Canadian Association of Broadcasters.................	½	-
Canadian Broadcasting Corporation...................	½	-
Canadian Telecommunications Carriers Association......	4	4
Téléglobe Canada (ex-COTC)...........................	1	3
Telesat Canada.......................................	½	-
CHILI - *CHILE* - CHILE		
Compañía de Teléfonos de Chile.......................	-	½
Empresa Nacional de Telecomunicaciones S.A. (ENTEL Chile)	½	½
COTE D'IVOIRE - *IVORY COAST* - COSTA DE MARFIL		
Société de télécommunications internationales de la Côte d'Ivoire (INTELCI).............................	-	½

*From ITU, Adminstrative Council, 35th Session, Geneva, May 1980, Document No. 5454-E (CA 35-22), 28 February 1980, pp. 119-137.

Note: Recognized private operating agencies and scientific or industrial organizations are classified in French alphabetical order of the members which approved their request to participate.

Participation in the work of a consultative committee is shown by the number of units corresponding to the class of contribution chosen, and non-participation, by the sign-.

Nom - *Name* - Nombre	CCIR	CCITT
DANEMARK - *DENMARK* - DINAMARCA		
Grande compagnie des télégraphes du Nord............	–	½
ESPAGNE - *SPAIN* - ESPAÑA		
Compañía Telefónica Nacional de España..............	1	1
ETATS-UNIS - *UNITED STATES* - ESTADOS UNIDOS		
Aeronautical Radio, Inc.	½	½
American Telephone and Telegraph Co.	5	5
Communications Satellite Corporation................	1	1 ½
FTC Communications, Inc.	–	½
Graphnet, Inc.	–	½
GTE Service Corporation.............................	½	½
Hawaiian Telephone Company..........................	–	½
ITT World Communications, Inc.	1	1
RCA Corporation.....................................	1	–
RCA Global Communications, Inc.	–	1
Satellite Business Systems (SBS)...................	1	–
Telenet Communications Corporation.................	–	½
TRT Telecommunications Corporation.................	–	½
Tymshare, Inc.	–	½
Western Union International, Inc.	–	2
Western Union Telegraph Company.....................	1	1
PORTO RICO - *PUERTO RICO* - PUERTO RICO		
All America Cables and Radio, Inc. - Puerto Rico....	½[1]	½
FINLANDE - *FINLAND* - FINLANDIA		
The Finnish Broadcasting Company....................	½	–

[1] A dénoncé sa participation aux travaux du CCIR ; cette dénonciation prendra effet au 17 avril 1980 - *Has denounced its participation to the work of the CCIR ; this denunciation will take effect on 17 April 1980* - Ha denunciado su participación a los trabajos del CCIR ; esta denuncia surtirá efecto el 17 de abril de 1980.

Nom – _Name_ – Nombre	CCIR	CCITT
FRANCE – _FRANCE_ – FRANCIA		
Compagnie française de câbles sous-marins et de radio (France Câbles et Radio).........................	½	½
Thomson-CSF...	½	½
ITALIE – _ITALY_ – ITALIA		
Italcable, Servizi Cablografici, Radiotelegrafici e Radioelettrici, Società per Azioni................	½	½
RAI – Radiotelevisione Italiana.....................	½	½
SIP – Società Italiana per l'Esercizio Telefonico p.a	½	½
Telespazio S.p.A. per le Comunicazioni Spaziali......	½	½
JAMAIQUE – _JAMAICA_ – JAMAICA		
Jamaica International Telecommunications Limited (JAMINTEL)...	–	½
The Jamaica Telephone Company Limited...............	–	½
JAPON – _JAPAN_ – JAPÓN		
Kokusai Denshin Denwa Company, Ltd.	1	1
Nippon Hoso Kyokai (Broadcasting Corporation of Japan)	1	–
Nippon Minkan Hoso Renmei (National Association of Commercial Broadcasters in Japan).................	½	–
Nippon Telegraph and Telephone Public Corporation....	2	3
MEXIQUE – _MEXICO_ – MEXICO		
Teléfonos de México S.A.	–	½

Nom - *Name* - Nombre	CCIR	CCITT
NIGERIA		
Nigerian External Telecommunications Ltd.	½	½
NOUVELLE-ZELANDE - *NEW ZEALAND* - NUEVA ZELANDIA		
Broadcasting Council of New Zealand	½	-
PHILIPPINES - *PHILIPPINES* - FILIPINAS		
Philippine Communications Satellite Corporation......	-	½
Philippine Global Communications, Inc.	-	½
Philippine Long Distance Telephone Company, Inc.	½	½
PORTUGAL		
Companhia Portuguesa Rádio Marconi...................	½	½
ROYAUME-UNI - *UNITED KINGDOM* - REINO UNIDO		
British Broadcasting Corporation....................	½	½
Cable and Wireless Limited..........................	2	2
Independent Broadcasting Authority..................	½	-
International Marine Radio Company, Ltd.	½	½
Post Office du Royaume-Uni..........................	5	5
The Marconi International Marine Company, Ltd.	½	½
SUDAFRICAINE (République) - *SOUTH AFRICA (Republic of)* - SUDAFRICANA (República)		
South African Broadcasting Corporation..............	½	-

Nom - *Name* - Nombre	CCIR	CCITT
SUEDE - *SWEDEN* - SUECIA		
Swedish Broadcasting Corporation (Sveriges Radio)....	1	-
SUISSE - *SWITZERLAND* - SUIZA		
Radio-Suisse S.A. de télégraphie et téléphonie sans fil...	1	1
VENEZUELA		
Compañía Anónima Nacional Teléfonos de Venezuela.....	½	½
YOUGOSLAVIE - *YUGOSLAVIA* - YUGOSLAVIA		
Radiotélévision yougoslave.........................	½	-
Total	45 ½	49

II. Scientific or Industrial Organizations

Nom - *Name* - Nombre	CCIR	CCITT
ALLEMAGNE (République fédérale d') - *GERMANY (Federal Republic of)* - ALEMANIA (República Federal de)		
Allgemeine Elektricitäts-Gesellschaft, AEG-Telefunken Backnang	2	2
Felten & Guilleaume Carlswerk A.G., Köln-Mülheim	-	1
Kalle, filiale de Hoechst A.G., Wiesbaden	-	½
Rank Xerox G.m.b.H., Düsseldorf	-	½
Rohde & Schwarz, München	1	-
Siemens A.G., München	3	3
Standard Elektrik Lorenz A.G., Stuttgart	1	1
TE KA DE Felten & Guilleaume Fernmeldeanlagen G.m.b.H., Nürnberg	½	½
Telefonbau und Normalzeit Lehner & Co.,Frankfurt/Main	-	½
Wandel und Goltermann, Reutlingen	½	½
AUTRICHE - *AUSTRIA* - AUSTRIA		
Siemens Aktiengesellschaft Osterreich, Wien	-	½
BELGIQUE - *BELGIUM* - BELGICA		
Bell Telephone Manufacturing Company, Anvers	-	½
Câblerie de Charleroi, Charleroi	-	½
Codex Europe S.A., Bruxelles	-	½
GTE ATEA S.A., Herentals	-	½
Manufacture belge de lampes et de matériel électronique, S.A., Bruxelles	-	½
Manufactures de câbles électriques et de caoutchouc S.A., Eupen	-	½
Telindus, Bruxelles	-	½
CANADA - *CANADA* - CANADA		
Bell - Northern Research Ltd.,Ottawa (Ontario)	1	1
Spar Technology Limited, Ste-Anne-de-Bellevue (Québec)	½	-

Nom - *Name* - Nombre	CCIR	CCITT
ESPAGNE - *SPAIN* - ESPAÑA		
SECOINSA - Sociedad Española de Comunicaciones e Informática, S.A., Madrid......................	–	½
Standard Eléctrica S.A., Madrid......................	–	½
ETATS-UNIS - *UNITED STATES* - ESTADOS UNIDOS		
Anaconda Telecommunications, Anaheim (California)....	–	½ [2]
Arthur D. Little, Cambridge (Massachusetts)..........	–	½
Burroughs Corporation, Detroit (Michigan)...........	–	½
Codex Corporation, Newton (Massachusetts)...........	–	½
Compass Communications Corporation, Miami (Florida)..	–	½
Control Data Corporation, Greenwich (Connecticut)....	–	½
Databit, Inc., Hauppauge (New York).................	–	½
Extel Corporation, Northbrook (Illinois)............	–	½
Farinon Electric, San Carlos (California)..........	½	–
Ford Aerospace and Communications Corp., Washington	½	–
Frederick Electronics Corp. Frederick (Maryland).....	–	½
General Cable International, Inc.,Greenwich (Connecticut)	–	½
General DataComm Industries, Inc., Wilton (Connecticut)	–	½
General Electric Company, Information Services Business Division, Rockville (Maryland)............	½	½
GTE International Inc., Stamford (Connecticut).......	½	½
Harris Corporation, Satellite Communication Division, Melbourne (Florida)	½	½
Hewlett-Packard, Mountain View (California)..........	–	½
Honeywell, Inc., St. Petersburg (Florida)...........	–	½
Hughes Aircraft Company, Los Angeles (California)....	½	½
International Business Machines Corporation, Armonk (New York)..	–	½
International Telephone and Telegraph Corporation, (ITT), New York....................................	½	1
Intertel, Inc., Burlington (Massachusetts)..........	–	½
MCI Communications Corporation, Washington..........	–	½ [1]
Minnesota Mining and Manufacturing Company, St.Paul (Minnesota).......................................	–	½

[1] A dénoncé sa participation; cette dénonciation prendra effet au 15 janvier 1980 - *Has denounced its participation; this denunciation will take effect on 15 January 1980* - Ha denunciado su participación; esta denuncia surtirá efecto el 15 de enero de 1980.

[2] A dénoncé sa participation; cette dénonciation prendra effet au 29 octobre 1980 - *Has denounced its participation; this denunciation will take effect on 29 October 1980* - Ha denunciado su participación; esta denuncia surtirá efecto el 29 de octubre de 1980.

Nom - *Name* - Nombre	CCIR	CCITT
Etats-Unis - *United States* - Estados Unidos (suite - *continued* - continuación)		
National Scientific Laboratories, Division of Systematics General Corporation, Falls Church (Virg.)	½	–
Paradyne Corporation, Largo (Florida)	–	½
Racal-Milgo, Inc., Miami (Florida)	–	½
Rapicom, Inc., Fairfield (New Jersey)	–	½
Raychem Corporation, Menlo Park (California).........	–	½
Rixon Inc., Silver Spring (Maryland)	–	½
Rockwell International, Arlington (Virginia).........	½	½
Stromberg-Carlson Corporation, Longwood (Florida)....	–	½
Tele-Dynamics, Division of AMBAC Industries, Inc. Fort Washington (Pennsylvania)	–	½[1]
Tran Telecommunications Corporation, Marina del Rey (California)	–	½
United Telecommunications, Inc., Shawnee Mission (Kansas)...	–	½
Universal Data Systems, Huntsville (Alabama).........	–	½
Vadic Corporation, Sunnyvale (California)...........	–	½
Wang Laboratories, Inc., Lowell (Massachusetts)......	–	½
Western Electric International, Inc., Greensboro (North Carolina)	–	1
Xerox Corporation, Rochester (New York).............	–	½

1) A dénoncé sa participation; cette dénonciation prendra erfect au
   5 novembre 1980 - *Has denounced its participation; this denunciation
   will take effect on 5 November 1980* - Ha denunciado su participación;
   esta denuncia surtirá efecto el 5 de noviembre de 1980.

Nom - *Name* - Nombre	CCIR	CCITT
FINLANDE - *FINLAND* - FINLANDIA		
Oy Nokia Ab/Finnish Cable Works, Helsinki...........	-	½
Research Institute of the Helsinki Telephone Company, Helsinki	-	½
Telefenno Oy, Helsinki.............................	½	½
FRANCE - *FRANCE* - FRANCIA		
Association des Ouvriers en Instruments de Précision (A.O.I.P.), Paris................................	-	½
Compagnie générale de constructions téléphoniques,Paris	-	½
Compagnie Honeywell Bull, Paris.....................	-	½
Compagnie industrielle des télécommunications CIT-ALCATEL, Velizy-Villacoublay (Yvelines)...........	½	½
International Business Machines (IBM), Europe, Paris.	-	½
Le Matériel Téléphonique (L.M.T.) Boulogne-Billancourt	½	½
Les Câbles de Lyon, Lyon...........................	-	½
Lignes télégraphiques et téléphoniques (L.T.T.), Paris	-	½
Société anonyme de télécommunications, Paris.........	-	½
Société d'applications générales d'électricité et de mécanique (S.A.G.E.M.), Paris.....................	-	½
Société de constructions électromécaniques Jeumont-Schneider, Puteaux (Hauts-de-Seine)................	-	½
Société d'Etudes et de Constructions Electroniques (S.E.C.R.E.), Paris..............................	-	½
Société Enertec-Schlumberger, Saint-Etienne (Loire)..	-	½
Société française des téléphones Ericsson, Colombes..	-	½
Société Intertechnique, Orsay (Essonne).............	-	½
Société Kalle Infotec, Nanterre (Hauts-de-Seine).....	-	½
Société LEA, Laboratoire Electro-Acoustique, Rueil-Malmaison..................................	-	½
Société MATRA, Vélizy-Villacoublay (Yvelines)........	-	½
Société Sintra, Gennevilliers (Hauts-de-Seine).......	-	½
Société TEKELEC-AIRTRONIC, Sèvres (Hauts-de-Seine)...	-	½
Société "Télécommunications Radioélectriques et Téléphoniques" (T.R.T.), Paris....................	-	½
Société Texas Instruments, Villeneuve-Loubet (Alpes Maritimes)	-	½

1) A dénoncé sa participation aux travaux du CCIR ; cette dénonciation prendra effet au 29 janvier 1980 - *Has denounced its participation to the work of the CCIR ; this denunciation will take effect on 29 January 1980* - Ha denunciado su participación a los trabajos del CCIR ; esta denuncia surtirá efecto el 29 de enero de 1980.

Nom - *Name* - Nombre	CCIR	CCITT
SOFRECOM - Société française d'études et de réalisations d'équipements de télécommunications, Paris...	-	½
SOTELEC - Société mixte pour le développement de la technique des télécommunications sur câbles, Paris	-	½
Syndicat des industries téléphoniques et télégraphiques, Paris......................................	-	½
HONGRIE - *HUNGARY* - HUNGRIA		
Beloiannisz Híradástechnikai Gyár, Budapest..........	-	½
Institut de recherches des télécommunications (Távközlési Kutató Intézet), Budapest..............	½	-
ITALIE - *ITALY* - ITALIA		
AET - Applicazioni Elettro-Telefoniche, Torino.......	-	½
ANIE - Associazione Nazionale Industrie Elettrotecniche ed Elettroniche, Milano..................	½	-
CSELT - Centro Studi e Laboratori Telecomunicazioni, Torino...	½	½
FATME - Fabbrica Apparecchiature Telefoniche e Materiale Elettrico, Roma	-	½
GTE Telecomunicazioni S.p.A., Cassina de'Pecchi (Milano)...	½	½
Ing. C. Olivetti & Co., Direzione Marketing, Ivrea...	-	1
Philips S.p.A., Divisione Telecomunicazioni, Milano..	-	1
Pirelli S.p.A., Milano	½ 1)	½
Selenia, Roma	½	½
Sielte S.p.A. Impianti Elettrici e Telefonici Sistema Ericsson, Roma	-	½
Società Face Standard, Milano	-	½
Società Italiana Reti Telefoniche Interurbane (S.I.R.T.I.), Milano	½	½
Società Italiana Telecomunicazioni Siemens S.p.A., Milano	½	½
Società Marconi Italiana, Genova	½	½
Società Telettra, Milano	½	½
STET - Società Finanziaria Telefonica, Roma	½	½

Nom - *Name* - Nombre	CCIR	CCITT
JAPON - *JAPAN* - JAPON		
Communication Industries Association of Japan, Tokyo.	-	½
Communication Line Products Association of Japan, Tokyo	-	½
Electronic Industries Association of Japan, Tokyo....	½	-
NORVEGE -*NORWAY* - NORUEGA		
A/S Elektrisk Bureau, Nesbru	-	½
PAYS-BAS - *NETHERLANDS* - PAISES BAJOS		
N.V. Philips' Telecommunicatie Industrie, Hilversum..	1	2
ROYAUME-UNI - *UNITED KINGDOM* - REINO UNIDO		
BICC Telecommunication Cables Limited, Prescot (Merseyside).......................................	-	½
CASE - Computer and Systems Engineering Ltd., Rickmansworth (Herts)	-	½
Databit Ltd., Northampton	-	½
GEC Telecommunications Limited, Telephone Works, Coventry	-	½
Hewlett-Packard Limited, South Queensferry (West Lothian, Scotland)	-	½
International Aeradio Limited, Southall	-	½
International Computers Limited (ICL), London	-	½
Kalle Infotec Limited, London	-	½
Logica Limited, London	-	½
M.G.V.S. (Fax Communications) Co. Ltd., London	-	½
Muirhead Data Communications Limited, Beckenham	-	½
Plessey Company Limited, Maidenhead (Berkshire)......	½	½
Pye TMC Limited, Malmesbury (Wiltshire)	-	1[1)]
Racal-Milgo Limited, Reading Berks	-	½
Rank Xerox Limited, Mitcheldean (Gloucestershire)....	-	½
SE Labs (EMI) Ltd., Feltham (Middlesex)..............	-	½
Sperry Univac International Division, London	-	½
Standard Telephones and Cables Limited, London	-	½

1) A dénoncé sa participation; cette dénonciation prendra effet au 16 février 1980 - *Has denounced its participation; this denunciation will take effect on 16 February 1980* - Ha denunciado su participación; esta denuncia surtirá efecto el 16 de febrero de 1980.

Nom – *Name* – Nombre	CCIR	CCITT
SUEDE – *SWEDEN* – SUECIA		
ELLEMTEL Utvecklings Aktiebolag, Skärholmen	–	½
Telefonaktiebolaget L.M. Ericsson, Stockholm	1	2
SUISSE – *SWITZERLAND* – SUIZA		
Câbleries de Brugg, Cossonay et Cortaillod	–	½
Cerberus Ltd., Männedorf (Zürich)	–	½
Eurosat S.A., Genève	½	–
Gfeller S.A., Bern	–	½
Hasler S.A., Bern	–	1
Siemens – Albis S.A., Zürich	–	1
Sodeco-Saia S.A., Genève	–	½
Standard Téléphone et Radio S.A., Zürich	–	½
TCHECOSLOVAQUIE – *CZECHOSLOVAKIA* – CHECOSLOVAQUIA		
TESLA Electronics and Telecommunication, Prague	½	½
YOUGOSLAVIE – *YUGOSLAVIA* – YUGOSLAVIA		
ISKRA Electromehanika – Kranj, Kranj.................	½	½
NIKOLA TESLA – Telecommunication Equipment Factory, Zagreb	½	½
Total	25 ½	79 ½

III. *International Organizations*

Nom - *Name* - Nombre	CCIR	CCITT
Agence spatiale européenne - *European Space Agency* - Agencia Espacial Europea	½	-
Association des entreprises gouvernementales de télécommunications de l'Accord sous-régional andin - *Association of State telecommunication undertakings of the Andean Sub-regional Agreement* - Asociación de Empresas Estatales de telecomunicaciones del Acuerdo Subregional Andino (ASETA)...................	*)	*)
Association du transport aérien international - *International Air Transport Association (IATA)* - Asociación de Transporte Aéreo Internacional	*)	*)
Association européenne de constructeurs fabricants de calculateurs électroniques - *European Computer Manufacturers Association (ECMA)* - Asociación Europea Constructores Calculadoras Electrónicas ....:...	-	½
Association européenne des services informatiques - *European Computing Service Association* - Asociación Europea de Servicios de Informática (ECSA)	-	½
Association interaméricaine de radiodiffusion (AIR) - *Inter-American Association of Broadcasters (IAAB)* - Asociación Interamericana de Radiodifusión (AIR)......	*)	-
Association internationale de signalisation maritime (A.I.S.M.) - *International Association of Lighthouse Authorities (I.A.L.A.)* - Asociación Internacional de Señalización Marítima (A.I.S.M.).....................	½	-
Association internationale des usagers des télécommunications - *International Telecommunications Users Group* - Asociación Internacional de Usuarios de Telecomunicaciones (INTUG)	-	½
Bureau intergouvernemental pour l'informatique - *Intergovernmental Bureau for Informatics (IBI)* - Oficina Intergubernamental para la Informática	1)	1)

1) A présenté une demande d'exonération.

Nom - *Name* - Nombre	CCIR	CCITT
Bureau international de l'heure (BIH) - *International Time Bureau (ITB)* - Oficina Internacional de la Hora (OIH)...	*)	-
Chambre internationale de marine marchande - *International Chamber of Shipping* - Cámara Naviera Internacional (ICS)....................................	-	½
Comité de la recherche spatiale - *Committee on Space Research* - Comité de Investigación Espacial (COSPAR)..	*)	-
Comité international des télécommunications de presse (CITP) - *International Press Telecommunications Council (IPTC)* - Comité Internacional de las Telecomunicaciones de Prensa (CITP)	-	½
Comité international radio-maritime - *International Maritime Radio Association* - Comité Internacional Radiomarítimo (CIRM)...............................	*)	*)
Comité international spécial des perturbations radio-électriques - *International Special Committee on Radio Interference* - Comité Internacional Especial de Perturbaciones Radioeléctricas (CISPR)............	*)	-
Commission électrotechnique internationale (CEI) - *International Electrotechnical Commission (IEC)* - Comisión Electrotécnica Internacional (CEI).........	*)	*)
Commission inter-unions pour l'attribution de fréquences à la radioastronomie et à la science spatiale - *Inter-Union Commission on Frequencies Allocation for Radio Astronomy and Space Science* - Comisión Interuniones para la Atribución de Frecuencias a la Radioastronomía y la Ciencia Espacial (IUCAF)........	*)	-
Conférence internationale des grands réseaux électriques à haute tension - *International Conference on Large High Tension Electric Systems* - Conferencia Internacional de las Grandes Redes Eléctricas de Alta Tensión (CIGRE).....................................	*)	*)

Nom - *Name* - Nombre	CCIR	CCITT
Congrès international de télétrafic (CIT) - *International Teletraffic Congress (ITC)* - Congreso Internacional de Teletráfico (ITC)...............................	-	*)
Conseil international des unions scientifiques (CIUS) - *International Council of Scientific Unions (ICSU)* - Consejo Internacional de Uniones Científicas (CIUC)...	*)	-
Fédération internationale d'astronautique - *International Astronautical Federation* - Federación Astronáutica Internacional (IAF).................................	*)	-
Fédération internationale de documentation - *International Federation for Documentation* - Federación Internacional de Documentación (FID)......	*)	*)
Fédération internationale pour le traitement numérique de l'information - *International Federation for Information Processing* - Federación Internacional de Tramitación de Información (IFIP)	-	*)
Organisation arabe de télécommunications par satellite - *Arab Satellite Communications Organization* - Organización Arabe de Comunicaciones por Satélite (ARABSAT)	$\frac{1}{2}$	$\frac{1}{2}$
Organisation de la télévision ibéro-américaine - *Ibero-American Television Organization* - Organización de la Televisión Iberoamericana (OTI)	*)	*)
Organisation internationale de normalisation - *International Organization for Standardization* - Organización Internacional de Unificación de Normas (ISO)..	-	*)
Organisation internationale de police criminelle (OIPC) - *International Criminal Police Organization (ICPO)* - Organización Internacional de Policía Criminal (OIPC)	-	*)
Organisation internationale de radiodiffusion et télévision - *International Radio and Television Organization* - Organización Internacional de Radiodifusión y de Televisión (OIRT).............................	*)	*)

Nom - *Name* - Nombre	CCIR	CCITT
Organisation internationale de télécommunications par satellites - *International Telecommunications Satellite Organization* - Organización Internacional de Telecomunicaciones por satélite (INTELSAT)........	$\frac{1}{2}$	$\frac{1}{2}$
Union africaine des postes et télécommunications - *African Postal and Telecommunications Union* - Unión Africana de Correos y Telecomunicaciones (UAPT)......	*)	*)
Union arabe des télécommunications - *Arab Telecommunication Union* - Unión Arabe de Telecomunicaciones.....	*)	*)
Union astronomique internationale (UAI) - *International Astronomical Union (IAU)* - Unión Astronómica Internacional (UAI)..	*)	-
Union de radiodiffusion "Asie-Pacifique" - *Asia-Pacific Broadcasting Union (ABU)* - Unión de radiodifusión "Asia-Pacífico"	*)	-
Union de radiodiffusion des Etats arabes - *Arab States Broadcasting Union* - Unión de Radiodifusión de los Estados Arabes (ASBU)............................	*)	-
Union des radiodiffusions et télévisions nationales d'Afrique - *Union of National Radio and Television Organizations of Africa* - Unión de las Radiodifusiones y Televisiones Nacionales de Africa (URTNA)........	*)	*)
Union européenne de radiodiffusion (UER) - *European Broadcasting Union (EBU)* - Unión Europea de Radiodifusión (UER)......................................	*)	*)
Union internationale de l'industrie du gaz (UIIG) - *International Gas Union (IGU)* - Unión Internacional de la Industria del Gas (UIIG).....................	-	*)
Union internationale des chemins de fer - *International Union of Railways* - Unión Internacional de Ferrocarriles (UIC)..	-	*)

Nom - *Name* - Nombre	CCIR	CCITT
Union internationale des producteurs et distributeurs d'énergie électrique - *International Union of Producers and Distributors of Electrical Energy* - Unión Internacional de Productores y Distribuidores de Energía Eléctrica (UNIPEDE)................................	-	*)
Union internationale des radio-amateurs - *International Amateur Radio Union* - Unión Internacional de Aficionados de Radio (IARU)...............................	*)	-
Union internationale des transports publics - *International Union of Public Transport* - Unión Internacional de Transportes Públicos (UITP)...............	-	*)
Union panafricaine des télécommunications (UPAT) - *Panafrican Telecommunication Union (PATU)* - Unión Panafricana de Telecomunicaciones (UPAT)...........	*)	*)
Union radio-scientifique internationale - *International Union of Radio Science* - Unión Radiocientífica Internacional (URSI)...............................	*)	*)
Total	2	3 ½

*) Organisation exonérée de toute contribution aux dépenses, en application de la Résolution N°574 (modifiée) du Conseil d'administration - *Organization exempt from contribution to the expenses, under Administrative Council Resolution No. 574 (amended)* - Organización exonerada de toda participación en los gastos, en aplicación de la Resolución N.°574 (modificada) del Consejo de Administración.

	Nombre d'unités *Number of units* Número de unidades	
	CCIR	CCITT
Exploitations privées - *Private Operating Agencies* - Empresas privadas de explotación.........................	45 ½	49
Organismes scientifiques ou industriels - *Scientific or Industrial Organizations* - Organismos cientfficos o industriales...	25 ½	79 ½
Organisations internationales - *International Organizations* - Organizaciones internacionales....................	2	3 ½
Total	73	132

ANNEX F
Conversion Table: Swiss Francs to U.S. Dollars*

Year	Swiss Francs = $1.00
1946	4.280
1947	4.280
1948	4.280
1949	4.289
1950	4.316
1951	4.329
1952	4.312
1953	4.281
1954	4.282
1955	4.281
1956	4.280
1957	4.282
1958	4.282
1959	4.315
1960	4.313
1961	4.313
1962	4.315
1963	4.311
1964	4.308
1965	4.307
1966	4.316
1967	4.318
1968	4.305
1969	4.303
1970	4.295
1971	4.115
1972	3.793
1973	3.169
1974	2.955
1975	2.571
1976	2.492
1977	2.398
1978	1.786
1979	1.663
1980	1.674
1981	1.964
(January) 1982	1.842

*The authors and publisher wish to thank the *Banque Nationale Suisse* and the *Union De Banques Suisses* for supplying this information.

Selected Bibliography

This bibliography contains a selection of primary and secondary sources on the International Telecommunication Union and related subjects. Although some materials will be presented that antedate the Atlanctic City Conferences of 1947, primary emphasis will be on the post-1947 period. For an extensive bibliography of materials before 1947, consult Codding, *The International Telecommunication Union,* pp. 479-496.

I. Documents of the ITU and its Predecessors
A. *Telegraph and Telephone Conference Documents*
   (in chronological order)
1. *Conference telegraphique internationale de Paris, 1865.*
2. *Conference telegraphique internationale de Vienne, 1868.*
3. *Conference telegraphique internationale de Rome, 1871-1872.*
4. *Conference telegraphique internationale de St. Petersbourg, 1875.*
5. *Conference telegraphique internationale de Londres, 1879.*
6. *Conference telegraphique internationale de Berlin, 1885.*
7. *Conference telegraphique internationale de Paris, 1890.*
8. *Conference telegraphique internationale de Budapest, 1896.*
9. *Conference telegraphique internationale de Londres, 1903.*
10. *Conference telegraphique internationale de Lisbonne, 1908.*
11. *Conference telegraphique internationale de Paris, 1925.*
12. *Conference telegraphique internationale de Bruxelles, 1928.*
13. *Conference telegraphique internationale de Madrid, 1932.*
14. *Conference telegraphique et telephonique internationale de Caire, 1938.*
15. *International Telegraph and Telephone Conference, Paris, 1949.*
16. *Administrative Telegraph and Telephone Conference, Geneva, 1958.*
17. *World Administrative Telegraph and Telephone Conference, Geneva, 1973.*

B.   Radio Conference Documents
    (in chronological order)
1. *Conference preliminaire concernant la telegraphie sans fil, Berlin,
   1903.*
2. *Conference radiotelegraphique internationale de Berlin, 1906.*
3. *Conference radiotelegraphique internationale de Londres, 1912.*
4. *[Preliminary] International Conference on Electrical Communications, 1920.*
5. *Conference radiotelegraphique internationale de Washington,
   1927.*
6. *Conference radio-electrique europeene de Prague, 1929.*
7. *Conference radiotelegraphique internationale de Madrid, 1932.*
8. *Conference europeenne de radiodiffusion, Lucerne, 1933.*
9. *Conference internationale des radiocommunications du Caire,
   1938.*
10. *Conference europeenne des radiocommunications de Montreux
    (mars/avril, 1939).*
11. *International High Frequency Broadcasting Conference, Atlantic
    City, 1947.*
12. *International Radio Conference, Atlantic City, 1947.*
13. *European Regional Broadcasting Conference, Copenhagen, 1948.*
14. *European Regional Conference for the Maritime Mobile Radio
    Service, Copenhagen, 1948.*
15. *International Administrative Aeronautical Radio Conference,
    1st Session, Geneva, 1948.*
16. *International High Frequency Broadcasting Conference, Mexico
    City, 1948-1949.*
17. *Provisional Frequency Board, Geneva, 1948-1951.*
18. *Administrative Radio Conference for the African Region, Geneva,
    1949.*
19. *Administrative Radio Conference for Region 1, Geneva, 1949.*
20. *Administrative Radio Conference for Region 2 and the Fourth
    InterAmerican Radio Conference, Washington, D.C., 1949.*
21. *Administrative Radio Conference for Region 3, Geneva, 1949.*
22. *International Administrative Aeronautical Radio Conference,
    2nd Session, Geneva, 1949.*
23. *Special Administrative Conference for the Northeast Atlantic
    (Loran), Geneva, 1949.*
24. *Third North American Regional Broadcasting Conference, 1st
    Session, Montreal, 1949.*

25. *International High Frequency Broadcasting Conference, Florence/Rapallo, 1950.*
26. *Third North American Regional Broadcasting Conference, 2nd Session, Washington, D.C., 1950.*
27. *Extraordinary Administrative Radio Conference, Geneva, 1951.*
28. *European Broadcasting Conference, Stockholm, 1952.*
29. *Baltic and North Sea Radiotelephone Conference, Goteborg, 1955.*
30. *Ordinary Administrative Radio Conference, Geneva, 1959.*
31. *Special Regional Conference, Geneva, 1960.*
32. *European VHF/UHF Broadcasting Conference, Stockholm, 1961.*
33. *African VHF/UHF Broadcasting Conference, Geneva, 1963.*
34. *Extraordinary Administrative Radio Conference to Allocate Frequency Bands for Space Radiocommunications Purposes, Geneva, 1963.*
35. *African LF/MF Broadcasting Conference, Geneva, 1964.*
36. *Extraordinary Administrative Radio Conference for the Preparation of a Revised Allotment Plan for the Aeronautical (R) Service, 1st Session, Geneva, 1964.*
37. *African LF/MF Broadcasting Conference, Geneva, 1966.*
38. *Extraordinary Administrative Radio Conference for the Preparation of a Revised Allotment Plan for the Aeronautical Mobile (R) Service, 2nd Session, Geneva, 1966.*
39. *World Administrative Radio Conference to Deal with Matters Relating to the Maritime Mobile Service, Geneva, 1967.*
40. *World Administrative Radio Conference for Space Telecommunications, Geneva, 1971.*
41. *Regional Administrative Conference (Regions 1 and 3) to Draw up Frequency Assignment Plans for LF/MF Broadcasting, 1st Session, Geneva, 1974.*
42. *World Administrative Radio Conference for Maritime Mobile Telecommunications, Geneva, 1974.*
43. *Regional Administrative Radio Conference (Regions 1 and 3) to Draw up Frequency Assignment Plans for LF/MF Broadcasting, 2nd Session, Geneva, 1975.*
44. *World Administrative Radio Conference for the Planning of the Broadcasting-Satellite Service in Frequency Bands 11.7-12.2 GHz (in Regions 2 and 3) and 11.7-12.5 GHz (in Region 1), Geneva, 1977.*
45. *World Administrative Radio Conference on the Aeronautical Mobile (R) Service, Geneva, 1978.*

46. *World Administrative Radio Conference, Geneva, 1979.*
47. *Regional Administrative MF Broadcasting Conference (Region 2), 1st Session, Buenos Aires, 1980.*
48. *Regional Administrative MF Broadcasting Conference (Region 2), 2nd Session, Rio de Janeiro, 1981.*

C. Plenipotentiary Conference Documents
(in chronological order)

1. *Moscow Telecommunications Conference, Moscow, 1946.*
2. *International Telecommunication Conference, Atlantic City, 1947.*
3. *Plenipotentiary Conference, Buenos, Aires, 1952.*
4. *Plenipotentiary Conference, Geneva, 1959.*
5. *Plenipotentiary Conference, Montreux, 1965.*
6. *Plenipotentiary Conference, Malaga-Torremolinos, 1973.*

D. *Administrative Council Session Documents*

1. 2nd Session, Geneva, February, 1948.
2. 3rd Session, Geneva, September-October, 1948.
3. 4th Session, Geneva, August-September, 1949.
4. 5th Session, Geneva, September-October, 1950.
5. 6th Session, Geneva, April-May, 1951.
6. 7th Session, Geneva, April-June, 1952.
7. 8th Session, Geneva, May-June, 1953.
8. 9th Session, Geneva, May, 1954.
9. 10th Session, Geneva, April-May, 1955.
10. 11th Session, Geneva, April-May 1956.
11. 12th Session, Geneva, April-May 1957.
12. 13th Session, Geneva, April-May 1958.
13. 14th Session, Geneva, May-June, 1959.
14. 15th Session, Geneva, May-June, 1960.
15. 16th Session, Geneva, April-May, 1961.
16. 17th Session, Geneva, May-June, 1962.
17. 18th Session, Geneva, April-May, 1963.
18. 19th Session, Geneva, April-May, 1964.
19. 20th Session, Geneva, April-May, 1965.
20. 21st Session, Geneva, May-June, 1966.
21. 22nd Session, Geneva, May, 1967.
22. 23rd Session, Geneva, May, 1968.
23. 24th Session, Geneva, May, 1969.
24. 25th Session, Geneva, May-June, 1970.
25. 26th Session, Geneva, May, 1971.

26. 27th Session, Geneva, May-June, 1972.
27. 28th Session, Geneva, May-June, 1973.
28. 29th Session, Geneva, June-July, 1974.
29. 30th Session, Geneva, June, 1975.
30. 31st Session, Geneva, June, 1976.
31. 32nd Session, Geneva, May-June, 1977.
32. 33rd Session, Geneva, May-June, 1978.
33. 34th Session, Geneva, June, 1979.
34. 35th Session, Geneva, May, 1980.
35. 36th Session, Geneva, June, 1981.

E. Documents of CCIF Plenary Assemblies

1. First Plenary Assembly, Paris, April-May, 1924.
2. Second Plenary Assembly, Paris, June, 1925.
3. Third Plenary Assembly, Paris, November-December, 1926.
4. Fourth Plenary Assembly, Como, September, 1927.
5. Fifth Plenary Assembly, Paris, June, 1928.
6. Sixth Plenary Assembly, Berlin, June, 1929.
7. Seventh Plenary Assembly, Brussels, June, 1930.
8. Eighth Plenary Assembly, Paris, September, 1931.
9. Ninth Plenary Assembly, Madrid, September, 1932.
10. Xth Plenary Assembly, Budapest, September, 1934.
11. XIth Plenary Assembly, Copenhagen, June, 1936.
12. XIIth Plenary Assembly, Cairo, February, 1938.
13. XIIIth Plenary Assembly, London, October, 1945.
14. XIVth Plenary Assembly, Montreux, October, 1946.
15. XVth Plenary Assembly, Paris, July, 1949.
16. XVIth Plenary Assembly, Florence, October, 1951.
17. XVIIth Plenary Assembly, Geneva, October, 1954.
18. XVIIIth Plenary Assembly, Geneva, December, 1956.

F. Documents of CCIR Plenary Assemblies

1. First Meeting, The Hague, September-October, 1929.
2. Second Meeting, Copenhagen, May-June, 1931.
3. Third Meeting, Lisbon, September-October, 1934.
4. Fourth Meeting, Bucharest, May-June, 1937.
5. Vth Plenary Assembly, Stockholm, July, 1948.
6. VIth Plenary Assembly, Geneva, June-July, 1951.
7. VIIth Plenary Assembly, London, September-October, 1953.
8. VIIIth Plenary Assembly, Warsaw, August-September, 1955.

9.   IXth Plenary Assembly, Los Angeles, April, 1959.
10. Xth Plenary Assembly, Geneva, January-February, 1963.
11. XIth Plenary Assembly, Oslo, June-July, 1966.
12. XIIth Plenary Assembly, New Delhli, January-February, 1970.
13. XIIIth Plenary Assembly, Geneva, July, 1974.
14. XIVth Plenary Assembly, Kyoto, June, 1978.
15. XVth Plenary Assembly, Geneva, February, 1982.

G.   Documents of CCIT Plenary Assemblies

1.   First Plenary Assembly, Berlin, November, 1926.
2.   Second Plenary Assembly, Berlin, June, 1929.
3.   Third Plenary Assembly, Berlin, May, 1931.
4.   Fourth Plenary Assembly, Prague, May-June, 1934.
5.   Fifth Plenary Assembly, Warsaw, October, 1936.
6.   Sixth Plenary Assembly, Brussels, May, 1949.
7.   Seventh Plenary Assembly, Arnhem, June, 1953.
8.   Eighth Plenary Assembly, Geneva, December, 1956.

H. Documents of CCITT Plenary Assemblies

1.   Ist Plenary Assembly, Geneva, December, 1956.
2.   IInd Plenary Assembly, New Delhi, November-December, 1960.
3.   IIIrd Plenary Assembly, Geneva, June, 1964.
4.   IVth Plenary Assembly, Mar del Plata, October, 1968.
5.   Vth Plenary Assembly, Geneva, December, 1972.
6.   VIth Plenary Assembly, Geneva, September-October, 1976.
7.   VIIth Plenary Assembly, Geneva, November, 1980.

II. Other ITU Documents

1. Administrative Council, *Financial Regulations,* Edition 1979, Geneva, 1979.
2. Administrative Council, *Report on the Activities of the International Telecommunication Union,* (annual).
3. Administrative Council, *Resolutions and Decision of the International Telecommunication Union* (General revision, 1979).
4. Administrative Council, *Rules of Procedure of the Administrative Council (1976 Revision),* Geneva, 1976.
5. CCIR, *Circular Letters.*
6. CCIR, *Technical Bases for the World Administrative Conference, 1979,* Geneva, 1978.
7. CCITT, *Circular Letters.*
8. CCITT, *Economic and Technical Aspects of the Choice of Telephone Switching Systems,* Geneva, 1981.
9. *Final Acts of the Extraordinary Administrative Radio Conference, Geneva, 1951,* Geneva, 1952.
10. *Final Acts of the Extraordinary Administrative Radio Conference to Allocate Frequency Bands for Space Radiocommunication Purposes, Geneva, 1963,* Geneva, 1963.
11. *Final Acts of the International Telecommunication and Radio Conferences, Atlantic City, 1947,* Atlantic City, 1947.
12. *Final Acts of the World Administrative Radio Conference, Geneva, 1979,* Vols. I and II, Geneva, 1979.
13. *Final Acts of the World Administrative Radio Conference for the Planning of the Broadcasting-Satellite Service in Frequency Bands 11.7-12.2 GHz (in Region 1), (Geneva, 1977),* Geneva, 1977.
14. *Final Acts of the World Administrative Radio Conference for Space Telecommunications (Geneva, 1971),* Geneva, 1971.
15. *Final Acts of the World Administrative Telegraph and Telephone Conference (Geneva, 1973),* Geneva, 1973.
16. General Secretariat, *Circular Letters.*
17. General Secretariat, *Staff Regulations and Staff Rules of the International Telecommunication Union,* Geneva, 1969 (updated to 1 January 1980).
18. *Geneva: World Telecommunication Exhibition (Telecom 75),* Geneva 1976.
19a. IFRB, *Circular Letters.*
19b. IFRB, *Frequency Management and the Use of the Radio Frequency Spectrum and the Geostationary Satellite Orbit,* Geneva, 1979 (mimeo).

20. IFRB, *High Frequency Broadcasting Schedule* (quarterly).
21. IFRB, *International Frequency List.*
22. IFRB, *Minutes of Meetings.*
23. IFRB, *Rules of Procedure.*
24. IFRB, *Technical Standards.*
25. *International Telecommunication Convention, (Madrid, 1932),* Berne, 1933.
26. *International Telcommunication Convention, Atlantic City, 1947,* Berne, 1947.
27. *International Telecommunication Convention, Buenos Aires, 1952,* Geneva, 1953.
28. *International Telecommunication Convention, Geneva, 1959,* Geneva, 1960.
29. *International Telecommunication Convention, Montreux, 1965,* Geneva, 1966.
30. *International Telecommunication Convention, Malaga-Torremolinos, 1973,* Geneva, 1974.
31. *Pan-African Telecommunication Network,* Geneva, 1979.
32. *Radio Regulations* (1959), Geneva, 1960.
    Radio Regulations, Edition of 1976, Vols. I and II, Geneva, 1976.
34. *Radio Regulations, Edition of 1982,*Vols. I and II, Geneva, 1982.
35. Study Group for the Preparation of a Draft Constitutional Charter, *Report of the Study Group Appointed to Prepare a Draft Constitutional Charter,* Document No. SGC/39-E, 5 March 1969.
36. *Telecommunication Journal (Journal telegraphique* from 1868 to 1933.
37. *Yearbook of Common Carrier Telecommunication Statistics,* 8th edition, 1980.

III. Other Government Documents

1. France, Ministere des Travaux Publics, Sous-Secretariat d'Etat des Postes et des Telegraphes, *Conference internationale pour l'amilioration des communications postales et ferroviaires, telegraphiques et telephoniques et radiotelegraphiques (7-13 juillet 1920)*, Paris, 1920.

2. France, Sous-Secretariat d'Etat des Postes, Telegraphes et Telephones, *Comite technique de radiocommunication internationales — Etats-Unis, France, Grande-Bretagne, Italie, Japon, Paris: juin, juillet, aout 1921*, Paris, 1922.

3. France, Office of the Minister of War, Cipher Section, *EU-F-GB-I Protocol of 25 August 1919*, Paris, 1919.

4. Technical Committee on International Radio-Communication, *Report*, Paris, 1921.

5. UN, General Assembly, Official Records, Twenty-Fourth Session, *Review of the Administrative and Management Procedures Concerning the Programme and Budget of the International Telecommunication Union*, New York, 1970.

6. Universal Electrical Communications Union, *Draft of Convention and Regulations*, Washington, D.C., 1920.

7. U.S., Congress of the United States, Office of Technology Assesment, *Radiofrequency Use and Management: Impacts of the World Administrative Radio Conference of 1979*, Washington, D.C., 1982.

8. U.S., Department of State, *Arrangement effected by exchange of notes between the United States, Canada, Cuba and Newfoundland relative to the Assignment of High Frequencies to Radio Stations on the North American Continent*, Treaty Series No. 777-A, Washington, D.C., 1929.

9. U.S., Department of State, *EU-F-GB-I Radio Protocol*, Washington, D.C., Government Printing Office, 1920.

10. U.S., Department of State, Office of International Communications Policy, *Report of the Chairman of the United States Delegation to the World Administrative Radio Conference of the International Telecommunication Union, Geneva, Switzerland, September 24-December 8, 1979*, TD Serial No. 116, Washington, D.C., 1980 (mimeo).

11. U.S., Department of State, Office of International Telecom-
     munications Policy, *Report of the United States Representa-
     tive on the Administrative Council of the International Tele-
     communication Union, 31st Session, Geneva, Switzerland,
     June 14-July 2, 1976,* TD Serial No. 72, Washington, D.C., 1976
     (mimeo).
12. U.S., Department of State, Office of International Communi-
     cations Policy, *Report of the United States Representative on
     the Administrative Council of the International Telecomuni-
     cation Union, 32nd Session, Geneva, Switzerland, May 23-
     June 10, 1977,* TD Serial No. 86, Washington, D.C., 1977
     (mimeo).
13. U.S., Department of State, Office of Telecommunications,
     *Report of the United States Delegation to the Plenipotentiary
     Conference of the International Telecommunication Union,
     Montreux, Switzerland, September 14 to November 12, 1965,*
     TD Serial No. 973 Washington, D.C., 1965, (mimeo).
14. U.S., Department of State, Office of Telecommunications,
     *Report of the United States Delegation to the Plenipotentiary
     Conference of the International Telecommunication Union,
     Malaga-Torremolinos, Spain, September 14-October 15, 1973,*
     TD Serial No. 43, Washington, D.C., 1973, (mimeo).
15. U.S., Department of State, Office of Telecommunications,
     *Report of the United States Delegation to the World Maritime
     Administrative Radio Conference of the International Tele-
     communication Union, Geneva, Switzerland, April 22-June 7,
     1974,* TD Serial No. 50, Washington, D.C., 1974 (mimeo).
16. U.S., Department of State, Telecommunications Division, *Re-
     port of the Chairman of the United States Delegation to the
     Plenipotentiary Conference of the International Telecommun-
     ication Union, Geneva, Switzerland, October-December, 1959,*
     TD Serial No. 906, Washington, D.C., 1960 (mimeo).
17. U.S., Department of State, Telecommunications Policy Staff,
     *Report of the United States Delegation to the Plenipotentiary
     Conference of the International Telecommunication Union,
     Atlantic City, 1947,* TD Serial No. 341, Washington, D.C., 1947
     (mimeo).

18. U.S., Department of State, U.S. Mission in Geneva to Secretary of State, International Telecommunication Union: 34th Session of the Administrative Council, (Geneva, June 4-22, 1979) Situation Report, Incoming Telegrams Nos. 09823, 10256, and 10564, Washington, D.C., 1979.

19. U.S., Department of State, U.S. Mission in Geneva to Secretary of State, International Telecommunication Union: 35th Session of the Administrative Council, Situation Report, Incoming Telegram Nos. 06995, 07141, 07408, 07587, 07632, and 07768, Washington, D.C., 1980.

20. U.S., Department of State, U.S. Mission in Geneva to Secretary of State, International Telecommunication Union: 36th Session of the Administrative Council (Geneva, 1-19 June 1981) Incoming Telegram, Nos. 05650, 05761, 06148, 06223, and 06268, June, 1981.

21. U.S., Department of State, Telecommunications Division, *Report of the Chairman of the United States Delegation to the Administrative Telegraph and Telephone Conference of the International Telecommunication Union, Geneva, Switzerland, September 29 through November 29, 1958,* TD Serial No. 893 Washington, D.C., 1959 (mimeo).

22. U.S., 94th Congress, 1st Session, United States Senate, Committee on Government Operations, *U.S. Participation in International Organization, Feb. 1977,* U.S. Government Printing Office, Washington, D.C., 1977.

IV. Books

1. Abshire, David M., *International Broadcasting: A New Dimension of Western Diplomacy,* Beverly Hills: Sage, 1976.

2. Alexandrowicz, Charles H., *The Law of Global Communications* New York: Columbia University Press, 1971.

3. Browne, Donald R., *International Radiobroadcasting: The Limits of the Limitless Medium,* New York: Praeger Publishers, 1982.

4. Codding, George A., Jr., *Broadcasting Without Barriers,* Paris: UNESCO, 1959.

5. Codding, George A., Jr., *The International Telecommunication Union, An Experiment in International Cooperation,* Leiden: E. J. Brill, 1952. Republished in 1972 by Arno Press, New York.

6. Codding, George A., Jr., *The Universal Postal Union,* New York: New York University Press, 1964.

7. Colliard, C. A., and A. C. Kiss, *Les telecommunications par satellite — Aspects juridiques,* Paris, 1970.
8. Crane, Rhonda J., *The Politics of International Standards: France and the Color T. V. War,* Norwood, N.Y.: Ablex Publishing Company, 1979.
9. Debbash, Charles, *Le Droit de la radio et de la television,* Paris: Presses Universitaires de France, 1969.
10. Emery, Walter B., *National and International Systems of Broadcasting,* East Lansing: Michigan State University Press, 1969.
11. Ganley, Oswald H., and Gladys D. Ganley, *The U.S. and its Communications and Information Resources: International Implications,* Cambridge: Harvard University Center for Information Policy Research, 1981.
12. Garmier, Jacques, *L'UIT et les telecommunications par satellites,* Brussels: Emile Bruylant, S.A., 1975.
13. Gerbner, George (ed.), *Mass Media Policies in Changing Cultures,* New York: John Wiley and Sons, 1977.
14. Harlow, Alvin F., *Old Wires and New Waves,* New York: D. Appleton Century, 1936.
15. Haschke, Herbert and Willi Paubel, *Der Internationale Fernmeldeverein: Die Vereinten Nationen und ihre Spezial Organisationen Dokumente,* Berlin: Staatsverlag der Deutschen Demokratischen Republik, 1977.
16. Herring, James M., and Gerald C. Gross, *Telecommunications, Economics and Regulation,* New York: McGraw Hill, 1936.
17. Hudson, Heather E., Douglas Goldschmidt, Edwin B. Parker and Andrew Hardy, *The Role of Telecommunications in Socio-Economic Development; A Review of the Literature with Guidelines for Further Investigation,* Keewatin Communications, May 1979.
18. Leive, David M., *International Telecommunications and International Law: The Regulation of the Radio Spectrum,* Leiden: A. W. Sijthoff, 1970.
19. Levin, Harvey, J., *The Invisible Resource: Use and Regulation of the Radio Spectrum,* Baltimore: John Hopkins University Press, 1971.
20. McAnay, Emile and others (eds.) *Structure and Communication,* New York: Praeger Publishers, 1981.
21. McWhinney, Edward (ed.), *The International Law of Communications,* Dobbs Ferry: Oceana Publications, 1971.
22. Michaelis, Anthony R., *From Semaphore to Satellite,* Geneva: ITU, 1965.

23. Namurois, Albert, *Structures and Organization of Broadcasting in the Framework of Radiocommunications,* EBU Monograph No. 8. Geneva: European Broadcasting Union, 1972.

24. Nicotera, F., *The Studies of the International Radio Consultative Committee,* Rome: Edizioni Moderne, 1955.

25. Nordenstreng, Kaarle and Herbert I. Schiller, *National Sovereignty and International Communication,* Norwood, N.J.: Ablex Publishing Corporation, 1979.

26. Pelton, Joseph W., *Global Communications Satellite Policy: Intelsat, Politics and Functionalism,* Mt. Airy, Md.: Lomond Books, 1974.

27. Pool, Ethiel de Sola (ed.), *Handbook of Communication,* Chicago: Rand McNally, 1973.

28. Queeney, K.M., *Direct Broadcasting Satellites and the United Nations,* Leiden; Sijthoff Noordhoff, 1978.

29. Robinson, Glen O., *Communications for Tomorrow: Policy Perspectives,* New York: Praeger Publishers, 1978.

30. Signitzer, Benno, *Regulation of Direct Broadcasting Satellites,* New York: Praeger Publishers, 1976.

31. Smith, Delbert D., *International Telecommunication Control: International Law and the Ordering of Satellite and Other Forms of International Broadcasting,* Leiden: A. W. Sijthoff, 1969.

32. Tomlinson, John D., *The International Control of Radiocommunications,* Geneva: Imprimerie de Geneve, 1938.

33. Wallenstein, Gerd D., *International Telecommunication Agreements,* Dobbs Ferry: Oceana Publications, Inc., 1977.

V. Articles and Chapters in Books

1. Abdel-Ghani, A.H., "The Role of the United Nations in the Field of Space Communications," *Telecommunication Journal,* Vol. 38, No. 5, (May, 1971), pp. 393-398.

2. "Activities of the International Frequency Registration Board," *Telecommunication Journal,* Vol. 40, No. 8 (August, 1973), pp. 402-407.

3. Aiya, S. V. C., "The Rural Telecommunication Problem of Developing Countries, *Telecommunication Journal,* Vol. 47, No. 2 (February, 1980), pp. 107-111.

4. Asghar, M. Malek, Y. Senuma, and R. Pinez, "Work of the World Plan Committee for the Development of Telecommunications, Paris, 1980, and the Evolution of Telephone Traffic," *Telecommunication Journal,* Vol. 47, No. 9 (September, 1980), pp. 557-571.

5. "Backstage Politics at CCITT," *IPTC Newsletter,* No. 36 (February, 1977), pp. 17-18.
6. Baczko, Henryk, "The Maintenance of Telecommunication Networks — A Crucial Problem of the Transport and Communications Decade for Africa," *Telecommunication Journal,* Vol. 47, No. 10 (October, 1980), pp. 634-636.
   Baczko, Henryk, "Results of the 23rd Session of the ITU Administrative Council," *Telecommunication Journal* Vol. 35, No. 7 (July, 1968), pp. 291-293.
7. Baillod, Remy, "Problem es de mesures en telecommunications," in *La Mesure dans les telecommunications — Measurement in Telecommunications,* Paris: Centre National d'Etudes des Telecommunications, 1977, pp. 9-14.
8. Bardoux, Michel, "Geographical Distribution and Career Services in the ITU," *Telecommunication Journal,* Vol. 45, No. 5 (May, 1978), pp. 217-218.
9. Bazzy, William, "Telecommunication, Trade and the ITU," *Telecommunications,* Vol. 9, No. 4 (April, 1975), p. 8; and No. 5 (May, 1975), p. 34.
10. Ben Cheikh, Habib, "Role of the Development Bank," *Telecommunication Journal,* Vol. 47, No. 3 (March, 1980), pp. 157-158.
11. Berrada, Abderrazak, "The Background to Planning," *Intermedia,* Vol. IV (December, 1976), p. 9.
12. Berrada, Abderrazak, "The Effect of the 1979 WARC on the ITU," *Intermedia,* Vol. 8, No. 6 (November, 1980), pp. 32-33.
13. Berrada, Abderrazak, "Role of the International Frequency Registration Board," *Telecommunications,* Vol. 15, No. 6 (June, 1981), pp. 58-59.
14. Berrada, Abderrazak, "The 1971 Conference and Afterwards," *Intermedia,* Vol. IV, (December, 1976), pp. 10-11.
15. Branscomb, Anne W., "Making WARC Work," *Foreign Policy,* No. 34 (Spring, 1979), pp. 139-148.
16. Branscomb, Lewis M., "Computer Technology and the Evolution of World Communications," *Telecommunication Journal,* Vol. 47, No. 4 (April, 1980), pp 206-210.
17. Burtz, Leon, "The CCITT and the Development of Telecommunication Services," *Telecommunication Journal,* Vol. 45, No. 2 (February, 1978), p. 56.

18. Burtz, Leon, "Switching and the CCITT," *Telecommunication Journal,* Vol. 46, No. 4 (April, 1979), pp. 191-192.
   Burtz, Leon, "CCITT VIIth Plenary Assembly," *Telecommunication Journal,* Vol. 48, No. IV (April, 1981), pp. 166-167.
19. Busak, Jan, "The Geostationary Satellite Orbit — International Cooperation or National Sovereignty?" *Telecommunication Journal,* Vol. 45, No. 4 (April, 1978), pp. 167-170.
20. Busak, Jan, "On the Eve of the World Administrative Radio Conference for Space Telecommunications," *Telecommunication Journal,* Vol. 38, No. 2 (February, 1971), pp. 96-100.
21. Busak, Jan, "Some Reflections on Application Satellites," *Telecommunication Journal,* Vol. 43, No. 5 (May, 1976), pp. 355-360.
22. Butler, Richard E., "Allocating Satellite Frequencies," *Intermedia* Vol. IV (June, 1976), pp. 22-23.
23. Butler, Richard E., "The Importance of Cooperation in International Telcommunications," *Telephony,* Vol. 193 (November 7, 1977), pp. 51-52, 54, 56, 59, 61, 64, 66, 103, 104, and 108.
24. Butler, Richard E., *"Direct Broadcasting Satellites as a Factor in the Development of International Telecommunication Policy — The ITU World Administrative Radio Conference for the Planning of the Broadcasting-Satellite Service,"* Telecommunication *Journal,* Vol. 43, No. 4 (April, 1976), pp. 300-306.
25. Butler, Richard E., "Some Perspectives and Future Trends in Planning, Execution and Management of Technical Cooperation," *Telecommunication Journal,* Vol. 40, No. 8 (August, 1973), pp. 525-531.
26. Butler, Richard E., "World Administrative Radio Conference for Planning Broadcasting Satellite Service," *Journal of Space Law,* Vol. 5, No. 1 and 1 (Spring and Fall, 1977), pp. 93-99.
27. Chapuis, Robert, "The CCIF and The Development of International Telephony (1923-1956)," *Telecommunication Journal,* Vol. 43, No. 4 (March, 1976), pp. 184-197.
28. Chapuis, Robert, "History of Regulations Governing the International Telephone Service," *Telecommunication Journal,* Vol. 43, No. 3 (March, 1976), pp. 203-205.
29. Chapuis, Robert, "The Role of an International Telecommunication Consultative Committee in Technical Assistance," *Telecommunication Journal,* Vol. 24, No. 2 (February, 1957), pp. 32-36.
30. Chapuis, Robert, "Work of the Plan Committee in the Intercontinental Sphere (Rome, December, 1963)," *Telecommunication Journal,* Vol. 31, No. 4 (April, 1964), pp. 98-115.

31. Chassignol, Andre, "L'UIT apres Malaga-Torremolinos," *Revue francaise des telecommunciations,* No. 11 (April, 1974), pp. 57-61 and No. 12 (July, 1974), pp. 11-15.
32. Chayes, Abram and Leonard Chazen, "Policy Problems in Direct Broadcasting From Satellites," *Stanford Journal of International Studies,* Vol. V (June, 1970), pp. 4-20.
33. Codding, George A., Jr., "Agenda: Nairobi, 1982," *Telecommunications,* Vol. 15, No. 7 (July, 1981), pp. 59-61.
34. Codding, George A., Jr. "Influence in International Conferences: A Research Note," *International Organization,* Vol. 35, No. 4 (Fall 1981), pp. 715-724.
35. Codding, George A., Jr., "International Constraints on the Use of Telecommunications: The Role of the International Telecommunication Union," in L. Lewin (ed.), *Telecommunications: An Interdisciplinary Survey,* Dedham: Artech House, Inc., 1979.
36. Codding, George A., Jr., "The ITLU in the 1980s," *Intermedia,* Vol. 8, No. 5 (September, 1980), pp. 12-17.
37. Codding, George A., Jr., "The New Nations and the ITU: Some Policy Implications for the Future," in Herbert S. Dordick (ed.), *Proceedings of the Sixth Annual Telecommunications Policy Research Conference* Boston: D. C. Heath and Company, 1978, pp. 357-374.
38. Codding, George A., Jr., "The United States and the ITU in a Changing World," *Telecommunication Journal,* Vol. 44, No. 5 (May, 1977), pp. 231-235.
    Codding, George A., Jr., "The World Radio Conference and the Developing Countries," *Toward Freedom,* Vol. XXX, No. 1 (January, 1981), pp. 2-4.
40. Cotten, Charles-Herve, "Action internationale," *Revue francasie des telecommunications,* No. 17 (October, 1975), pp. 10-12.
41. Croze, Raymond J., "CCITT: VIth Plenary Assembly," *Telecommunication Journal,* Vol. 43, No. 12 (December, 1976), pp. 730-733.
42. de Wolf, Francis Colt, "The ITU and Global Communications," *Transaction of the Institute of Radio Engineers, Professional Group on Communication Systems,* Vol. CS-2, No. 3 (November, 1954).
43. Doran-Veevers, D. V., "ITU and the Information Era," *Telecommunications,* Vol. 15, No. 8 (August, 1981), pp. 54 and 62.

44. Dormer, D. J., "The CCITT and its Secretariat," *Telecommunication Journal,* Vol. 32, No. 8 (August, 1965), pp. 312-318.
45. Engvall, Lars, "International Network Planning — Two Case Studies: The Pan-African Telecommunication Network, and the Middle East and Mediterranean Telecommunication Network" *Telecommunication Journal,* Vol. 44, No. 7 (July, 1977), pp. 331-337.
46. Esping, E., "European Conference for the Examination of the Television and Broadcasting Frequency Bands Between 41 and 216 Mc/s," *Telecommunication Journal,* Vol. 20, No. 8 (August, 1953), pp. 120-121.
47. Fackelman, Mary P., International Telecommunications Bibliograph, Department of Commerce, (Office of Telecommunications Policy, 1976) (OT Special Publication 76-7).
48. Fernandez-Shaw, Felix, "The New International Telecommunication Convention (ITC) of Malaga-Torremolinos (1973)," *EBU Review,* Vol. 25, No. 2 (March, 1974), pp. 21-26.
49. Fijalkowski, W. J., "A Further Contribution to the Discussion on the Structure of the ITU," *Telecommunication Journal,* Vol. 32, No. 5 (May, 1965), pp. 204-212.
50. Ford, Michael, "The Search for International Standards," *Intermedia,* Vol. 7, No. 3 (May, 1979), pp. 48-50.
51. "Former CCITT Chief Answers IPTC Newsletter Critic," *IPTC Newsletter,* No. 36 (February, 1977), pp. 2-3.
52. "Frequency Regulation," *Intermedia,* Vol. IV (December, 1976), pp. 23-24.
53. Gayer, John, "Past and Future — Integrating the ITU Headquarters," *Telecommunication Journal,* Vol. 30, No. 6 (June, 1964), pp. 159-165.
54. Glazer, J., Henry "Inflex ITU — The Need for Space Age Revisions to the International Telecommunications Convention," *Federal Bar Journal,* Vol. 23, No. 1(Winter 1963), pp. 1-36.
55. Glazer, J. Henry, "The Law-Making Treaties of the International Telecommunication Union Through Time and in Space," *Michigan Law Review,* Vol. 60, No. 3(January, 1962), pp. 269-316.
56. Glazier, J. Henry, "Some Interpretive Grapeshot Concerning the Application of the International Telecommunication Convention to Military Radio Installations," *Federal Bar Journal,* Vol. 25, No. 3 (Summer, 1965), pp. 307-314.

57. Gould, Richard G. and Edward E. Reinhart, "The 1977 WARC on Broadcasting Satellites: Spectrum Management Aspects and Implications," *IEEE Transactions on Electromagnetic Compatibility,* Vol. EMC-19, No. 3 (August, 1977), pp. 171-178.
58. Goy, Raymond, "La repartition des frequences en matiere de telecommunications," *Americaine francaise de droit internationale,* Vol. V (1959), pp. 569-590.
59. Gross, Gerald C., "The New ITU — A Plan for the Reorganization of the Union," *Telecommunication Journal,* Vol. 30, No. 10 (October, 1963), pp. 305-311.
60. Gross, Gerald C., "Technical Assistance in the International Telecommunication Union," *Telecommunication Journal,* Vol. 21, No. 9 (September, 1954), pp. 150-159.
61. Gross, Gerald C., "Towards the Streamlining of the ITU Convention," *Telecommunication Journal,* Vol. 32, No. 2 (February, 1965), pp. 56-71.
62. Gross, Leo, "International Law Aspects of the Freedom of Information and the Right to Communicate," in Philip C. Horton (ed.), *The Third World and Press Freedom,* New York: Praeger Publishers, 1978.
    Groszkowski, Janusz, "The Origins of the CCIR — The Historical Background of Radio Science and Technology," *Telcommunication Journal,* Vol. 45, No. 6 (June, 1978), pp. 276-278.
63. Gosztony, G., K. Rahko and R. Chapuis, "The Grade of Service in the Worldwide Telephone Network," *Telecommunication Journal,* Vol. 46, No. 10 (October, 1979), pp. 627-633.
64. Harbi, M., "LF and MF Broadcasting Prospects for the Forthcoming ITU Conference, Geneva, 1974," *Telecommunication Journal,* Vol. 40, No. 11 (November, 1973), pp. 716-720.
65. Herrera-Esteban, Leon, "The Malaga-Torremolinos Conference," *Telecommunication Journal,* Vol. 41, No. 5 (May, 1974), pp. 294-298.
66. Hinchman, Walter, R., "The Technological Environment for International Communications Law," in Edward McWhinney (ed.), *The International Law of Communications* Leiden: A. W. Sijthoff, 1971.
67. Honig, David E., "Lessons for the 1999 WARC," *Journal of Communication,* Vol. 30, No. 2 (Spring, 1980) pp. 48-58.
68. Hough, Richard R., "Strategies for Dealing with an Integrated World Network," *Telecommunication Journal,* Vol. 47, No. 5 (May, 1980), pp. 259-262.

69. Hummel, E., "The Role of the CCITT in the Introduction of New Telecommunication Servies," *Telecommunication Journal,* Vol. 47, No. 1 (January, 1980), pp. 26-27.
70. Hummel, E., "State of CCITT Standardization on Public Data Networks," *Telecommunication Journal,* Vol. 46, No. 1 (January, 1979), pp. 33-39.
71. Ickowitz, Allan H., "The Role of the International Telecommunication Union in the Settlement of Harmful Interference Disputes," *Columbia Journal of Transnational Law,* Vol. 13, No. 1 (1974), pp. 82-97.
72. "International Cooperation Efforts in the Field of Telecommunications: Current Situation and Problems," *Look Japan,* Vol. 20 (May 10, 1975), pp. 14, 16 and 20.
73. "The International Telegraph and Telephone Consultative Committee and Technical Cooperation," *Telecommunication Journal,* Vol. 40, No. 8 (August, 1973), pp. 416-423.
74. "Introduction to the Concept of the Development of Human Resources by Means of Technical Assistance," *Telecommunication Journal,* Vol. 40, No. 8 (August, 1973), pp. 516-520.
75. "ITU's Role in the Telecommunication Development in ESCAP Region," *Look Japan,* Vol. 20 (April 10, 1975), p. 12.
77. "ITU's Technical Assistance to New and Developing Countries Under the UNDP and Other Related Aid Programmes," *Telecommunication Journal,* Vol. 40, No. 8 (August, 1973), pp. 455-463.
78. Jackson, C., "The Allocation of the Radio Spectrum," *Scientific American,* February, 1980, pp. 34-39.
79. Jacobson, Harold K., "The International Telecommunication Union: ITU's Structure and Functions," *Global Communications in the Space Age: Toward a New ITU,* Appendix C, New York: John and Mary Markle Foundation and the Twentieth Century Fund, 1973, pp. 59-101.
80. Jacobson, Harold K., "International Institutions for Telecommunication: The ITU's Role," in Edward McWhinney (ed.), *The International Law of Communications,* Leiden: A. W. Sijthoff, 1971.
81. Jacobson, Harold K., "ITU: A Potpourri of Bureaucrats and Industrialists," in Robert W. Cox and Harold K. Jacobson (eds.), *The Anatomy of Influence: Decision Making in International Organization* (New Haven: Yale University Press, 1974) pp. 59-101.

82. Jones, W. T., "The Role of International Standardization in Tele-
    communication Network Development," *Telecommunication
    Journal,* Vol. 40, No. 4 (April, 1973), pp. 207-210.
83. Keenleyside, H. L., "Administrative Problems of the United
    Nations Technical Assistance Administration," *Telecommuni-
    cation Journal,* Vol. 23, No. 5 (May, 1956), pp. 114-121, and No. 6
    (June, 1956), pp. 147-156.
84. Khabiri, Mohamad K., "International Frequency Registration
    Board (IFRB) Assistance to Member Countries of the ITU in
    Matters Relating to Frequency Management," *IEEE Transac-
    tions on Electromagnetic Compatibility,* Vol. EMC-19, No. 3
    (August, 1977), pp. 179-182.
85. Kirby, Richard C., "CCIR XIVth Plenary Assembly," *Telecom-
    munication Journal,* Vol. 45, No. 12 (December, 1978), pp. 643-648.
86. Kirby, Richard C., "Fifty Years of the International Radio Con-
    sultative Committee (CCIR)," *Telecommunication Journal,* Vol.
    45, No. 6 (June, 1978), pp. 267-275.
87. Kirby, Richard C., "Getting the Message Across: the Role of the
    International Consultative Committees in Global Communica-
    tions," *Telecommunications,* Vol. 15, No. 3 (March, 1981), pp.
    27-30.
88. Kitahara, Yasusada, "New Telecommunications in the Informa-
    tion Society," *Telecommunication Journal,* Vol. 47, No. 7 (July,
    1980), pp. 449.
89. Kopitz, D., "Considerations on the Regional Administrations
    LF/MF Broadcasting Conference (Second Session)" *EBU Review,*
    No. 155 (February, 1976), pp. 2-11.
90. Lalou, J., "The CCITT and the Development of Telephony Since
    1956," *Telecommunication Journal,* Vol. 45, No. 3 (March, 1976),
    pp. 198-202.
91. Levin, Harvey J., "Orbit and Spectrum Resource Strategies —
    Third World Demands," *Telecommunications Policy, Vol. 5, No. 2*
    (June, 1981) pp. 102-110.
92. Levin, Harvey J., "The Radio Spectrum Resource," *Journal of
    Law and Economics* Vol. XI (October, 1968), pp. 433-501.
93. Lieve, David M., "Regulating the Use of the Radio Spectrum,"
    Stanford Journal of International Studies, Vol. V (June, 1970),
    pp. 21-52.
94. Lonberg, Ib., "The Broadcasting Satellite Conference," *Tele-
    communication Journal,* Vol. 44, No. 10 (October, 1977), pp.
    482-488.

95. Matsumoto, Takashi, "Some Suggestions on CCIT Meeting," *Telecommunication Journal,* Vol. 37, No. 11 (November, 1970), p. 749.

96. McWhinney, Edward, "The Development of an International Law of Communications," in Edward McWhinney (ed.), *The International Law of Communications* Leiden: A. W. Sijthoff, 1971.

97. Miles, Paul D., "International Radio Frequency Management," *Telecommunication Journal,* Vol. 21, No. 9 (September, 1954), pp. 170-173.

98. Mill, Mohamed, "The CCITT — A Tradition of Method and Continuity," *Telecommmunication Journal,* Vol. 40, No. 2 (February, 1973), pp. 64-67.

99. Mili, Mohamed, "Frequency Regulation," *Intermedia,* Vol. IV (December, 1976), pp. 23-24.

100. Mili, Mohamed, "From Assistance to Cooperation," *Telecommunication Journal,* Vol. 40, No. 8 (August, 1973), pp. 390-391.

101. Mili, Mohamed, "Importance of Telecommunications for Developing Countries," *The Indian and Eastern Engineer,* 119th Anniversary Edition, as quoted in ITU, *Teleclippings,* No. 525, 29 mai 1978, p. 5.

102. Mili, Mohamed, "The Institutional Framework of Technical Cooperation in the ITU," *Telecommunication Journal,* Vol. 46, No. 8 (August, 1979), pp. 466-469.

103. Mili, Mohamed, "International Jurisdiction in Telecommunications Matters," *Telecommunication Journal,* Vol. 40, No. 3 (March, 1973), pp. 122-128, No. 4 (April, 1973), pp. 174-182, No. 6 (June, 1973), pp. 286-290, No. 7 (July, 1973), pp. 344-348, No. 9 (September, 1973), pp. 562-566, No. 12 (December, 1973), pp. 746-749; and Vol. 41, No. 3 (March, 1974), pp. 170-172.

104. Mili, Mohamed, "Linking Communications Across National Boundaries," *Telephony,* Vol. 191 (July 5, 1976), pp. 46-47.

105. Mili, Mohamed, "The New Regulations Governing Maritime Regulations," *Telecommunication Journal,* Vol. 41, No. 5, (May, 1974), pp. 342-346.

Mili, Mohamed, "The Regional Unions and the ITU," *Telecommunication Journal,* Vol. 44, No. 9 (September, 1977), p. 406.

106. Millington, George, "The CCIR and the Ionosphere," *The Radio and Electronic Engineer,* Vol. 45 (January-February, 1975), pp. 42-46.

107. Minaslan, Jora R., "Property Rights in Radiation: An Alternative Approach to Radio Frequency Allocation," *Journal of Law and Economics,* Vol. 18, No. 1 (April, 1975), pp. 221-72.
108. Moreira da Silva, Mario, "The International Law on Radio Interference," *EBU Review (Part B),* Vol. 73 (1962), pp. 37.
109. Moylan, James J., "The Role of the International Telecommunication Union for the Promotion of Peace Through Communications Satellites," *Case Western Reserve Journal of International Law,* Vol. 4, No. 1 (Winter, 1971), pp. 61-78.
110. Mulatier, Leon, "The International Telecommunication Union since 1947," *Telecommunication Journal,* Vol. 20, No. 12 (December, 1953), pp. 190-193.
111. Nicoteria, Federico, "The Structure of the ITU," *Telecommunication Journal,* Vol. 31, No. 6 (June, 1964), pp. 157-158.
112. O'Brien, Rita Cruse and G. K. Helleiner, "The Political Economy of Information in a Changing International Order," *International Organization,* Vol. 34, No. 4 (Autumn, 1980), pp. 445-470.
113. O'Neill, John J., Jr., "Technical Assistance to Developing Countries," *Telecommunications,* Vol. 15, No. 4 (April, 1981), pp. 67-68, 70 and 79.
114. O'Neill, John J., Jr., "The International Telecommunication Union," *Telecommunications,* Vol. 15, No. 2 (Feruary, 1981), pp. 25-26 and 97.
115. Perret, Robert-Louis, "International Law in Space — The Regulation of Satellite Telecommunications," *INTERAVIA,* Vol. 30 (December, 1975), pp. 1256-1258.
116. Persin, J., "Some Reflections on Technical Assistance," *Telecommunication Journal,* Vol. 23, No. 3 (March, 1956), pp. 50-53.
117. Persin, J., "Whither the ITU in Space?" *Telecommunication Journal,* Vol. 29, No. 3 (March, 1962), pp. 85-86.
118. Pierce, William B. and Nicolas Jequier, "The Contribution of Telecommunications to Economic Development," *Telecommunications Journal,* Vol. 44, No. 11 (November, 1977), pp. 532-534.
119. Pierre, E., "CCITT Activities in the Field of Rural Telecommunications," *Telecommunication Journal,* Vol. 47, No. 2 (February, 1980), pp. 112-115.
120. Ploman, Edward W., "Linking Broadcast Satellites in the UN and ITU," *Intermedia,* Vol. 5, (June, 1977), pp. 26-27.

121. Ploman, Edward W., "The Whys and Wherefores of International Organizations," *Intermedia,* Vol. 8, No. 4 (July, 1980), pp. 6-11.
122. Pool, Ithiel de Sola, Philip Stone, Alexander Szalzi, *Communications, Computers and Automation for Development,* New York: UNITAR, 1971.
123. Pool, Ithiel del Sola, "The Problems of WARC," *Journal of Communication,* Vol. 29, No. 1 (winter, 1979), pp. 187-196.
124. Poulequin, Herve, "The International Radio Consultative Committee (CCIR)," *Telecommunication Journal,* Vol. 33, No. 6 (June, 1966), pp. 224-234.
125. "Pre-investment Surveys," *Telecommunication Journal,* Vol. 40, No. 8 (August, 1973), pp. 505-515.
126. Rice, David M. "Regulation of Direct Broadcast Satellite: International Constraints and Domestic Options," *New York Law School Law Review,* Vo. XXV, No. 4 (1980), pp. 813-862.
127. Rose, D. C., "Regional Administrative LF/MF Broadcasting Conference (Geneva, 1975)," *Telecommunication Journal,* Vol. 43, No. 5 (May, 1976), pp. 349-354.
128. Rothblatt, Martin A., "International Legal Norms Governing Development of the Orbit/Spectrum Resource," *Telecommunications Policy,* Vol. 5, No. 2 (June, 1981), pp. 63-83.
129. Rouvierre, J., "The IVth Plenary Assembly of the CCITT," *Telecommunication Journal,* Vol. 35, No. 12 (December, 1968), pp. 631-637.
130. Rutkowski, Anthony, "Six Ad-Hoc Two: The Third World Speaks Its Mind," *Satellite Communications,* Vol. 4 (March, 1980), pp. 22-27.
    Rutkowski, Anthony M., "The 1979 World Administrative Radio Conference: The ITU in a Changing World," *The International Lawyer,* Vol. 13, No. 2 (Spring, 1979), pp. 289-327.
131. Rutkowski, Jerzy, "CCIR Studies on Measurements in Radiocommunications," *La Mesure dans les telecommunications,* Paris: Centre National d'Etudes des Telecommunications, 1977, pp. 6-8.
132 Rutkowski, Jerzy, "CISPR Plenary Assembly (Tokyo, July 1980), *Telecommunication Journal,* Vol. 47, No. 12 (December, 1980), pp. 749-751.
133. Rutman, Elaine, "CCIR et CCITT ou l'art de normaliser," *L'Onde Electrique,* Vol. 56 (November, 1976), pp. 485-486.

134. Ruud, H., "ITU Participation in the UNDP Today," *Telecommunication Journal,* Vol. 40, No. 8 (August, 1973), pp. 426-432.
135. Sathar, S. A., "The International Consultative Committees and the Problems of the New and Developing Countries," *Telecommunication Journal,* Vol. 31, No. 1 (January, 1964), pp. 21-23.
136. Sathar, S. A., "The International Consultative Committees and the Problems of the New and Developing Countries — A Progress Report," *Telecommunication Journal,* Vol. 35, No. 1 (January, 1968), pp. 16-20.
137. Sathar, S. A., "The International Consultative Committees and the Problems of New and Developing Countries — A Review of Progress," *Telecommunication Journal,* Vol. 38, No. 9 (September, 1971), pp. 642-645.
138. Segal, Brian, "International Negotiations on Telecommunications," *Intermedia,* Vol. 8, No. 6 (November, 1980), pp. 22-31.
139. Smith, Ernest K., "The History of the ITU with Particular Attention to the CCIT and CCIR and the Latter's Relations with URSI," *Radio Science,* Vol. II (June, 1976), pp. 497-507.
140. Spencer, G., "The Future of Medium- and Long-Wave Broadcasting," *Wireless World,* Vol. 80, No. 1464 (August, 1974), pp. 266-271.
141. Staudinger, Wilhelm, "Standardization Versus Innovation," *Telecommunication Journal,* Vol. 48, No. 3 (March, 1981), pp. 140-142.
142. Tar, Zoltan J., "CCITT: VIIth Plenary Assembly," *Telecommunication Journal,* Vol. 48, No. 4 (April, 1981), pp. 175-188.
143. "Technical Cooperation Activities in the Regions, Some Typical Features," *Telecommunication Journal,* Vol. 40, No. 8 (August, 1981), pp. 465-504.
144. "Technical Cooperation and the International Radio Consultative Committee," *Telecommunication Journal,* Vol. 40, No. 8 (August, 1973), pp. 408-415.
145. "Technical Cooperation in Telecommunications in 1966," *Telecommunication Journal,* Vol. 34, No. 7 (July, 1967), pp. 242-245.
146. "Telecom 79: 34th World Telecommunication Exhibition," *Telecommunication Journal,* Vol. 47, No. 2 (February, 1980), pp. 65-96.
147. Thompson, J. G., "The Dark Labyrinth of World Communication," *IPTC Newsletter,* No. 34 (May, 1976), pp. 1-5.

148. Thompson, J. G., "Get all Nations Involved in International Coordination," *Communication News,* Vol. 14 (January, 1977), p. 23.

149. Thompson, J. G., "International Communications 'Management' is 'Coordination'," *Communication News,* Vol. 12 (November, 1975), pp. 24-25.

150. Thorne, F. and F. J. Clarke, "World Administrative Radio Conference for Maritime Mobile Telecommunications (Geneva, 1974)," *Telecommunication Journal,* Vol. 41, No. 12 (December, 1974), pp. 733-740.
Thorne, F. and F.J. Clarke, *"ITU World Maritime Conference, Geneva, 1967,"* Telecommunication Journal, Vol. 35, No. 11 (February, 1968), pp. 58-65.

151. Toutan, Michel, "Planning in Telecommunications," *Telecommunication Journal,* Vol. 43, No. 3 (March, 1976), pp. 225-231.

152. "Union Activities in the Field of Technical Cooperation During the Period 1965-1972," *Telecommunication Journal,* Vol. 40, No. 8 (August, 1973), pp. 437-454.

153. "The Union's Participation in the United Nations Development System Up to 1964," *Telecommunication Journal,* Vol. 40, No. 8 (August, 1973), pp. 433-436.

154. Valensi, Georges, "Brief History of the CCIF," *Telecommunication Journal,* Vol. 41, No. 7 (July, 1974), pp. 416-417.

155. Valensi, Georges, "The Development of International Telephony — the Story of the International Telephone Consultative Committee, 1924-1956," *Telecommunication Journal,* Vol. 32, No. 1 (January, 1965), pp. 9-17.

156. Valters, Eric, N., "Perspectives in the Emerging Law of Satellite Communications," *Stanford Journal of International Studies,* Vol. V (June, 1970), pp. 53-83.

157. Verdon, Christiane, "L'UIT et quelques problemes existant au sein de cette organisation," *University of Toronto Law Journal,* Vol. XX, No. 3 (1970), pp. 386-388.

158. von Baeyer, Hans J., "Politics of Worldwide Telecommunications," *Telephony,* Vol. 195, No. 2 (July 10, 1978), pp. 97-100.

159. von Sanden, Dieter, "Public Global Telecommunication Network for Voice, Text, Picture and Data Transmission," *Telecommunication Journal,* Vol. 47, No. 4 (April, 1980), pp. 211-216.

160. Wallenstein, Gerd D., "Development of Policy in the ITU," *Telecommunications Policy,* Vol. 1, No. 2 (March, 1977).

161. Wallenstein, Gerd D., "Global Scope at the National Level: The ITU's Contribution to Telecommunications Development," *IEEE Transactions on Communications,* Vol. Com-24, No. 7 (July, 1976), pp. 700-709.

162. Wallenstein, Gerd., "Handbooks of the Consultative Committees: Bridges Between International Standardization and National Telecommunication Development," *Telecommunication Journal,* Vol. 43, No. 10 (October, 1976), pp. 633-637.

163. Wallenstein, Gerd D., "The Internationalization of Telecommunications Systems Development," *Telecommunication Journal,* Vol. 41, No. 1 (January, 1974), pp. 33-38.

164. Wallenstein, Gerd D., "Language Barriers to Technical Collaboration in the ITU," *Telecommunication Journal,* Vol. 41, No. 5 (May, 1974), pp. 313-317.

165. Wallenstein, Gerd D., "Make Room in Space — Harmony and Dissonance in International Telecommunications," *Telecommunication Journal,* Vol. 40, No. 1 (January, 1973), pp. 29-46; and No. 2 (February, 1973), pp. 95-102.

166. White, C. E., "Telecommunication Needs of Developing Countries," *Telecommunications,* Vol. 10 (September, 1976), pp. 49-53.

167. Wolter, Werner, "Participation on the ITU: A Two-way Street," *Telephony,* Vol. 194, No. 5 (January 30, 1978), pp. 82 and 84.

168. Woolley, Michael, "The Telephone, its Invention and Development," *Telecommunication Journal,* Vol. 34, No. 3 (March, 1976), pp. 175-183.

169. "World Administrative Radio Conference," *Telecommunication Journal,* Vol. 47, No. 1 (January, 1980), pp. 4-10.

170. "VIth Plenary Assembly of the CCITT," *Telecommunications,* Vol. 10, (June, 1976), pp. 199-226.

171. White, Curtis T., "Uprooting the Squatters," *Foreign Policy,* No. 34, (Spring, 1979), pp. 148-153.

172. "XIIth Plenary Assembly of the CCIR," *Telecommunication Journal,* Vol. 37, No. 5 (May, 1970), pp. 207-209.

173. "35 Session of the ITU Administrative Council," *Telecommunication Journal,* Vol. 47, No. 8 (August, 1980), pp. 494-495.

VI. Other

1. American Society of International Law, *The International Telecommunication Union: Issues and Next Steps.* Report by the Panel on International Telcommunications Policy of the American Society of International Law. New York: Carnegie Endowment for International Peace, 1971.

2. Butler, Richard A., "World Telecommunication Development and the Role of the International Telecommunication Union," speech delivered at the International Conference on Trans-National Data Regulation, Brussels, 9 February 1978.

3. Camara, Sajo Magan-Nyameh, "The Pan-African Telecommunications Network: An Analysis," (M.S. dissertation, University of Colorado, 1981).

4. Eugster, Ernest, "The Management of Television in Europe: The Experience of the EBU and OIRT." (These presentee a l'Universite de Geneve pour l'obtention du grade de Docteur es sciences politiques, Geneva, 1981).

5. Feldman, Mildred L. B., "The United States in the International Telecommunication Union and in Pre-ITU Conference," (Ph.D. dissertation, Louisiana State University, 1976).

6. Hudson, Heather (Study Director), *WARC 79: Development Communications Strategies: A Report to USAID,* report prepared for the Agency for International Development by the Academy for Educational Development, March, 1979.

7. Huszach, Frederick W., "The International Law-making Process: A Case Study on the International Regulation of Space Telecommunication," (Ph.D. dissertation, University of Chicago, 1964).

8. Lee, Robert E., "The ITU: Crosswinds of Change," remarks before the International Radio and Television Society, Inc., Americana Hotel, New York, November 13, 1974.

9. Rahim, Syed A., "ITU's Institutional Structure and Resource Allocation Issues," paper presented at the Advanced Seminar on International Communication Policy and Organizations, Communication Institute, East-West Center, Honolulu, July 5-August 3, 1979.

10. Richstad, Jim and Jackie Bowen, *International Communication Policy and Flow.* Honolulu: East-West Communication Institute, East-West Center, 1976.

11. Wheaton, Martha Jane, "The Preparation for and the Implications of the General World Administrative Radio Conference of 1979," (M.A. dissertation, Naval Postgraduate School, Monterey, California, 1975).
12. Woldman, Joel M., "The World Administrative Radio Conference of 1979: Preparations and Prospects," The Library of Congress, Congressional Research Service, Washington, D.C., June 29, 1979.

Abbreviations
and Acronyms

1. ABU   Asia-Pacific Broadcasting Union
2. AIR   *Asociacion Interamericana de Radiodifusion /* Inter-American Broadcaster Association
3. AM   amplitude modulation
4. APTU   African Postal and Telecommunication Union
5. ARABSAT   Arab Satellite Communications Organization
6. ASBU   Arab States Broadcasting Union
7. ASECNA   *Agence pour la Securite de la navigation aerienne en Afrique et Madagascar*
8. ASETA   *Associacion de Empresas Estatales de Telecommunicationes del Acuendo Subregional Andino /* Association of State Telecommunication Undertakings of the Andean Sub-regional Agreement
9. AT&T   American Telephone and Telegraph Company
10. ATU   Arab Telecommunication Union
11. BBC   British Broadcasting Company
12. CCIF   *Comite consultatif international telephonique /* International Telephone Consultative Committee 1934-1956. *Comite consultatif international des communications telephonique a grande distance /* International Consultative Committee on Long Distance Telephone Communication, 1924-1933.
13. CCIR   *Comite consultatif international des radiocommunications /* International Radio Consultative Committee
14. CCIT   *Comite consultatif international telegraphique /* International Telegraph Consultative Committee, 1925-1956

15. CCITT *Comite consultatif international telegraphique et tele-phonique*/International Telegraph and Telephone Consultative Committee

16. CEPT *Conference Europeene des Administrations des Postes et des Telecommunications*/European Conference of Postal and Telecommunications Administrations

17. CIGRE *Conference internationale des grands reseaux elec-triques a haute tension*/International Conference on Large High Tension Electric Systems

18. CIRM *Comite international radio-maritime*/International Maritime Radio Association

19. CISPR *Comite international special des perturbations radio-electriques*/International Special Committee on Radio Interference

20. COPUOS Committee on the Peaceful Uses of Outer Space (United Nations)

21. COSPAR Committee on Space Research

22. DBS Direct Broadcasting Satellite

23. EARC Extraordinary Administrative Radio Conference

24. EBU European Broadcasting Union

25. ECMA European Computer Manufacturers Association

26. ECSA European Computing Service Association

27. EEC European Economic Community

28. ESA European Space Agency

29. FID *Federation internationale de documentation*/International Federation for Documentation

30. FM frequency modulation

31. GHz gigahertz

32. HF high-frequency

33. IAAB Inter-American Association of Broadcasters

34. IAF International Astronautical Federation

35. IALA International Association of Lighthouse Authorities

36. IARU International Amateur Radio Union

37. IATA International Air Transport Association

38. IAU International Astronomical Union

39. IBI Intergovernmental Bureau for Informatics

40. ICAO International Civil Aviation Organization

41. ICPO International Criminal Police Organization (also known as INTERPOL)

42. ICS International Chamber of Shipping

43. ICRC International Committee of the Red Cross

44. ICSU International Council of Scientific Unions

45. IEC   International Electrotechnical Commission
46. IFIP   International Federation for Information Processing
47. IFL   International Frequency List
48. IFRB   International Frequency Registration Board
49. IGU   International Gas Union
50. IMCO   Intergovernmental Maritime Consultative Organization
51. IMF   International Monetary Fund
52. INTELSAT International Telecommunications Satellite Organization
53. INTERSPUTNIK International Space Telecommunication Organization
54. INTUG International Telecommunications Users Group
55. IPTC   International Press Telecommunications Council
56. ISM   Industrial, scientific, and medical applications
57. ISO   International Organization for Standardization (formerly International Standardization Organization)
58. ITB   International Time Bureau
59. ITC   International Teletraffic Congress
60. ITF   International Transport Workers Federation
61. ITU   International Telecommunication Union
62. IUCAF Inter-Union Commission on Frequency Allocations for Radio Astronomy and Space Sciences
63. kHz   kilohertz
64. LF   low frequency
65. MF   medium frequency
66. MHz   megahertz
67. NANBA North American National Broadcasters Association
68. NARBA North American Regional Broadcasting Agreement
69. OIRT   *Organisation internationale de radiodiffusion et television*/International Radio and Televison Organization
70. OTI   *Organizacion de la Television Iberoamericana*/Ibero-American Television Organization
71. PANAFTEL Pan-African Telecommunication Network
72. PATU   Panafrican Telecommunication Union
73. PFB   Provisional Frequency Board, 1949-1951.
74. PTT   Post, Telegraph and Telephone Administration [A term used to describe a ministry or other unit of government which provides these services].
75. RARC   Regional Administrative Radio Conference
76. RPOA   Recognized Private Operating Agency
77. SIO   Scientific or Industrial Organization

78. SDR   Special Drawing Right
79. UAPT   *Union africaine des postes et telecommunications /* African Postal and Telecommunications Union
80. UHF   ultra high frequency
81. UIC   *Union internationale des chemins de fer /* International Union of Railways
82. UIR   *Union international de radiodiffusion/* International Broadcasting Union (1926-1950)
83. *UITP* *Union internationale des transports publics /* International Union of Public Transport
84. UN   United Nations
85. UNDP   United Nations Development Program
86. UNESCO United Nations Educational, Scientific and Cultural Organization
87. UNIPEDE *Union internationale des producteurs et distributeurs d'energie electrique/* International Union of Producers and Distributors of Electrical Energy
88. UPU   Universal Postal Union
89. URSI   *Union radio-scientifique internationale/* International Union of Radio Science
90. URTNA   *Union des radiodiffusions et televisions nationales d'Afrique/* Union of National Radio and Television Organizations of Africa
91. VHF   very high frequency
92. WACC   World Association for Christian Communication
93. WARC   World Administrative Radio Conference
94. WIPO   World Intellectual Property Organization
95. WMO   World Meterorological Organization

Index